Association of Reptilian and Amphibian Veterinarians Conference 2017 (ARAV 2017)

Dallas, Texas, USA
23 – 28 September 2017

ISBN: 978-1-5108-5202-0

Printed from e-media with permission by:

Curran Associates, Inc.
57 Morehouse Lane
Red Hook, NY 12571

Some format issues inherent in the e-media version may also appear in this print version.

Printed by Curran Associates, Inc. (2018)

Association of Reptilian & Amphibian Veterinarians
5003 SW 41st Blvd
Gainesville FL 32608-4930
USA

Phone: 480-703-4941

info@arav.org

Additional copies of this publication are available from:

Curran Associates, Inc.
57 Morehouse Lane
Red Hook, NY 12571 USA
Phone: 845-758-0400
Fax: 845-758-2633
Email: curran@proceedings.com
Web: www.proceedings.com

Table of Contents

Section 1: Sunday Technician Session

Section 2: Plenary Session

Section 5: ARAV Nutrition

Section 6: ARAV Reproduction Bonus Session

Section 7: ARAV Diagnostics

Section 10: Honeybee Medicine

Section 11: ARAV Anesthesia

Section 12: ARAV Chelonians

Section 14: ARAV Master Classes

Section 21: ARAV Bonus Master Class

Section 22: ARAV Nutrition Bonus Session

Section 23: ARAV Reproduction

Section 24: ARAV Infectious Diseases

Section 25: ARAV Clinical Cases and Reviews

Section 1

Sunday Technician Session

Melissa Kling, DVM and
Louisa M. Asseo, DVM, DABVP (Canine and Feline Practice)
Session Chairs

Sponsored By:

Handling, Restraint, and Venipuncture of Exotic Animals

Teresa Bradley Bays, DVM, CVA, DABVP (Exotic Companion Mammal Practice)

Session #2021

Affiliation: From Belton Animal Clinic and Exotic Care Center, 1308 N Scott Avenue, Belton, MO 64012, USA and Raymore Animal Clinic, 406 W Pine Street, Raymore, MO 64083, USA.

Abstract: Methods of handling, restraint, and venipuncture in reptiles, small mammals and birds that are commonly seen in general practice are described and demonstrated. Defense mechanisms of each species are discussed along with zoonotic diseases, special handling circumstances and tips for caring for hospitalized exotics.

Basic Handling and Restraint

Reptiles and amphibians

Know the natural defense mechanisms of each species. Train staff members appropriately. Define roles prior to handling. Gather supplies prior to handling.

Lizards: tail whipping, biting, nails

Your defense: Control serpentine movement by holding the shoulders and pelvis while controlling the tail. Many lizards have the ability to "drop" their tails if restrained by the distal aspect. Be careful not to lean against the tail and trap it against the table as you restrain it.

Snakes: teeth, constriction, tail, defecation, urination, musk glands

Your defense: Support the snake well, hold head gently but firmly and move as the snake moves. Act like a tree a snake that is well supported is less likely to resist or get injured during handling and will be less stressed.

Locate the head. Pin the head. Support the entire body well (use multiple handlers if necessary). Place the tail first in the container.

Turtles and tortoises: nails, beak, legs, spurs and pointed shells, urine dumping (hold the patient out away from your shoes)

Your defense: Hold midway between front and rear legs. Sedation or anesthesia may be necessary for larger patients.

Amphibians: nets, plastic containers, restrain heads of hellbenders, large toads and giant salamanders, minimize handling, moistened exam gloves, moistened paper towel or dishcloth.

Ferrets

Most ferrets are easy to handle but a new ferret must be approached slowly until this can be verified. Allow ferret to approach you. Distract the ferret with one hand and then pick it up with the other hand by gently grasping it around the middle. Sleeping ferrets should be awakened before they are picked up.

If the ferret attempts to bite, keep it in motion. The scruff can be grasped if necessary; grab as deeply as you can from both sides of the neck. Distract it with your other hand or by blowing in its face. When a ferret bites it will usually not let go easily and if the victim struggles the ferret will bite harder. You may have to wait for it to let go. Placing it down on the table or floor, blowing in its face, or spraying it with water may get it to release.

Scruffing ferrets and allowing them to hang from the scruff virtually immobilizes them like a cat with a kitten. Sometimes the head needs to be pressed against the table to scruff them safely. Provide favored food items to facilitate exams.

Hamsters and gerbils

Hamsters and gerbils should be allowed to become fully awake before handling. They may be able to be scooped up by hand or in a container. They can be scruffed if necessary although this may be difficult to do and is not quite immobilizing like a cat with the kitten.

Gerbils can be lifted by the base of the tail. If lifted or restrained by the tip of the tail the skin of the tail will come off and create de-gloving injury. Provide favored food items to facilitate exams.

Rats and mice

Approach slowly. Allow to wake up if sleeping. It often helps to catch them in a container such as a coffee can or cardboard tube. Mice can be restrained by the base of the tail. Restraining by the tail away from the base will cause degloving of the skin to occur.

Rats can be restrained by the scruff but rarely need to be. Most rats when handled gently are easily handled. Provide favored food items to facilitate exams.

Sugar gliders

As a nocturnal species they are best handled at night or in dimmer light. They can be scruffed. They will "gritch" when wakened or frightened. They should be well supported when held. Gliders may jump or glide away and can be fast. They can be examined with little restraint if kept in a cloth, in their cloth pouch, or in the gram-scale bowl. Provide favored food items to facilitate exams and recovery from anesthesia and surgery.

Chinchillas

Chinchillas are normally gentle but not always used to being handled. They can bite if really frightened.

They will experience "fur slip" when restrained. Grasp gently but firmly by the base of the tail and support with the other hand behind the front legs. Provide favored food items to facilitate exams.

Guinea pigs

Talk to them quietly as you handle them. Scoop them up by placing one hand behind the front legs and one hand supporting the rump and rear legs (C position). They will readily walk off a table as they don't see well directly in front of them. They do best in pairs or trios but will fight if not bonded slowly.

Rabbits

Talk to them quietly as you handle them. Scoop them up by placing one hand behind the front legs and one hand supporting the rump and rear legs (C position). Hold under your arm like a football if frightened. Cover their eyes if necessary. They can break their back if kicking when not well restrained. They can break their backs and legs when jumping from your arms. They can break their backs when you place them in the cage as they jump in and kick back; therefore, place in backwards.

Birds

Minimize handling. Have everything ready before you handle. They need breast muscles to breathe, so do not constrict by holding too firmly over the keel area. Use a hand towel or paper towel. Dim lights and create a spot light. Hold the head between thumb and first finger, then wrap other fingers around wings

You can hold around cheeks or around top of the head and below beak. You can wrap your finger around the neck below the beak. Use fear-free techniques when possible, sedate or anesthetize if needed.

Venipuncture

Reptiles

Snakes: tail vein, ultrasound-guided cardiocentesis, jugular

Turtles: jugular, anterior brachial vein, tail vein

Lizards: jugular, tail vein, ventral abdominal vein

Small mammals

Ferrets: cranial vena cava, jugular, saphenous vein, cephalic vein

Hamsters and gerbils: cranial vena cava, orbital sinus bleed (?)

Rats and mice: cranial vena cava, ventral tail vein/artery, orbital sinus bleed (?)

Sugar gliders: cranial vena cava, saphenous vein, femoral vein

Chinchillas: cranial vena cava, cephalic vein, saphenous vein

Guinea pigs: cranial vena cava, jugular, cephalic vein, saphenous vein

Rabbits: lateral saphenous vein, cephalic vein, jugular vein

Anesthesia is recommended for all but rabbits and ferrets.

Birds

Avian species: right jugular vein, brachial vein, tarsometatarsal vein

Zoonotic Potential of Exotic Animals and Wildlife

Organisms with zoonotic potential include bacteria (*Yersinia, E. coli, Salmonella*), fungi and dermatophytes, *Mycobacterium*, and parasites (*Giardia*, roundworms [*Baylisascaris*], hookworms). Keep in mind the size of your patient when determining what procedures you will perform and what treatments you will choose. Keep in mind the level of patient compromise when determining procedure and treatment to be performed.

Care Tips for Exotic Animals

Know basic environmental needs. Consider temperature, temperature gradient, and humidity. Be prepared to stimulate the appropriate environment for each species when patients are hospitalized, boarded or impounded.

Rabbits and guinea pigs overheat readily. If kept outside, provide shade and ventilation. If not sure, provide a hide-box.

Know whether the species is carnivore, herbivore, or omnivore. Feed appropriately sized prey. The author prefers feeding dead prey. Never keep live prey in with the animal longer than 5-10 minutes.

Review of Intravenous Access in Reptiles and Amphibians

Ian Kanda, RVT

Session #2090

Affiliation: From Avian, Exotic and Zoological Service, Center for Veterinary Health Sciences, Oklahoma State University, Stillwater, OK 74078, USA.

Introduction

Reptiles and amphibians present many challenges with regard to gaining intravenous access and yet this is an essential skill, necessary for any technician working with exotic species. This is a review of blood collection as well as intravenous and intraosseous catheterization in snakes, lizards, turtles, and amphibians commonly seen in veterinary practice.

Lizards

Ventral tail vein

Primary access to the vasculature in lizards is the ventral tail vein. Venipuncture can be achieved in awake lizards, often with minimal restraint. Sedation or anesthesia should be considered in species that utilize tail autotomy as a defense mechanism. The vein runs ventral to the coccygeal vertebrae and can be accessed with either a ventral midline or a lateral approach. Ventral access is by inserting a needle upwards at an approximate 45° to 90° craniodorsal angle. Laterally, the needle is inserted into the midline, often a concave depression between the dorsal and ventral musculature. Once inserted, keeping minimal back-pressure on the syringe will allow a flash of blood into your needle to confirm placement, without collapsing the vessel. The ventral tail vein gets smaller as you approach more distal however avoid an approach that may hit the paired hemipenes or scent glands immediately caudal to the vent.

The ventral tail vein can be catheterized but it is important to remember that drugs injected into this vessel may pass through the renal portal system before the rest of the systemic circulation.[1,2] Access to this vein for catheterization can be had by bending the tail laterally. Insert the catheter in a cranial direction ventral to the transverse vertebral processes. Aim towards the midline of the tail. Advance the catheter off of the stylet immediately after penetration of the vein to avoid passing completely through.[2]

Cephalic vein

The cephalic vein runs along the dorsal surface of the front leg and is typically neither visible, nor palpable. There are no defining landmarks to aid in a blind stick attempt, therefore a cut down procedure is necessary to visualize the vein and is performed on the distal dorsal surface of the leg.[3] Local or general anesthesia is recommended.[4,5] Aseptically prepare the skin. The vein is superficial, make a small cut and then further locate it with blunt dissection.[1,4,5] Once located, insert the catheter into the vein in a proximal direction. Feed the catheter and remove the stylet. Attach an injection cap or T-port and close the skin with sutures or tissue glue.[1]

Jugular vein and cranial vena cava

In larger lizards the vasculature can be accessed using the jugular veins.[1,6] They generally cannot be visualized or palpated however the needle can be blindly inserted along a line caudal to the tympanic membrane down the jugular groove. Trans-illumination may be helpful in visualizing the jugular in some small lizards. Alternatively, the cranial vena cava may also be a point of vasculature access, especially in species that readily autotomize their tails.[7,8] Anesthesia is recommended.[8] The insertion site for the vena cava in mammals is next to the manubrium and the first rib.[9,10] In most lizards the heart is positioned very cranially, within the shoulder girdle. Due to this anatomy, access to the vena cava in lizards is further cranial on the ventral neck than when approached in mammals. This technique has not been fully described in reptiles.

Ventral abdominal vein

The ventral abdominal vein runs along the ventral midline of the coelom just below the linea alba.[1,3] It is accessed cranial to the umbilicus in the caudal coelom.[1] A needle is inserted along the ventral midline at a shallow, craniad angle. This venipuncture site is uncommonly used because applying adequate pressure can be difficult and hemorrhage is a common complication.[1]

Intraosseous catheterization

Many scenarios may require the placement of an intraosseous catheter. Sites for intraosseous catheterization in lizards include the femur, humerus and tibia.[1,4] Avoid IO catheters in animals with nutritional secondary hyperparathyroidism as iatrogenic fractures are likely. Chance of permanent disability, osteomyelitis and pain should all be considered prior to placement and IO catheters should not be in place longer than 72 hours.[1,4]

First, choose a site that is easily accessible, avoiding a bone if it is long and curved as this may make feeding a catheter difficult. In laterally compressed species, the proximal femur or humerus may be the choice access points. Adducting the leg inward will highlight the insertion points of the greater trochanter of the femur or the greater tubercle of the proximal humerus. This distal diaphysis of the femur or the proximal tibial crest can be accessed by flexing the stifle joint. Aseptically prepare the skin around the chosen site and inject lidocaine (2%) for local anesthesia.[4] Place a spinal needle firmly against the cortex of the bone and drill with rotation and gentle pressure. A stylet avoids catheter plugs but hypodermic needles without stylets can be used in smaller patients. Once the cortex has been punctured the catheter is now in the medullary cavity and can be carefully fed further in to the bone. Confirmation of placement includes the presence of bone marrow, a catheter that moves synonymously with the bone as the leg is moved, audible scraping of the catheter inside the bone as it is gently rotated and orthogonal radiographic views. Heparinized saline should flush easily and will insure patency. Place a small amount of antibiotic ointment on the insertion site to avoid infection and secure the catheter with tape wrapped around the leg both distally and proximally. A splint or additional bandaging can be used for extra protection as needed.

Snakes

Ventral tail vein

As with lizards, the ventral tail vein is often the primary venipuncture site in snakes. Insert the needle midline in between the ventral tail scales at a 45° to 90° angle. A good grasp of the tail is necessary to maintain midline alignment as snakes are good at twisting away. Avoid the hemipenes and scent glands that are immediately caudal to the vent by accessing the caudal 75% of the tail.[6] Slight back pressure on the syringe is maintained as the needle is inserted towards the vertebrae until a flash of blood indicates venipuncture.

Intra-cardiac

In snakes the heart beating can often be visually observed on the ventral surface. The heart is located at approximately the one third mark of the snout to vent length and an ultrasonic Doppler can be used to confirm the location.[1] Sedation is recommended. Palpate and immobilize the heart between two fingers and insert the needle in a craniodorsal direction into the ventricle using an approximate 45° angle immediately caudal to the heart. Insert in between two ventral scutes. For blood collection the syringe may fill slowly with each heartbeat. Avoid excess negative pressure on the plunger.[10] In emergency situations an intravenous catheter can be inserted into the ventricle using this technique. Upon withdrawal, maintain digital pressure on the site for 60 seconds.

Jugular vein

The jugular veins run ventromedial to the ends of the ribs. The right jugular is often larger but both can be used.[1,4] To locate the jugular, begin by locating the heart and counting 4-9 ventral scutes cranially.[3,11] Due to the prevalence of other collection sites, the jugular is uncommonly used, however it can be approached blindly by inserting the catheter at a shallow angle in a cranial direction lateral to the ventral scute and medial to the ribs. Sedation or anesthesia is recommended.[4,5] A cut down procedure greatly aids in the visualization of the vein and insertion of a catheter. Begin with aseptic preparation. Make a small cut and then further locate the vein with blunt dissection.[1,4,5] Insert the catheter in a proximal direction. Feed the catheter and then remove the stylet. Secure an injection cap or T-port and close the skin with sutures or tissue glue.[1]

Chelonians

Subcarapacial venous sinus

The subcarapacial venous sinus is typically the most accessible site for venipuncture in chelonians because it can be accessed while they are retracted into their shell. The sinus is found midline underneath the carapace and dorsal to the neck. To collect blood from this site, insert the needle midline where the skin meets the shell above the head. Continue insertion in a caudodorsal direction. With slight negative pressure, observe for blood in the syringe. Lymph contamination of blood samples is common because the subcarapacial sinus is caudal to large lymphatic vessels.[12] Do not redirect the needle while deep in the tissues or laceration of one of the many vessels in the area could occur.

Jugular vein

The jugular vein is an excellent venipuncture site in chelonians if you can access the neck. This can be easy in ill, sedate or cooperative turtles. The head can be grasped between the thumb and forefinger and gently extended from the shell. Some species may have a right jugular vein that is larger than the left.[6] The vein may be visible and can be occluded with digital pressure at the base of the neck. The jugular vein runs from the angle of the mandible to the carapacial inlet. Insert the needle or catheter at a shallow angle proximal to the tympanum in a caudal direction. A cut-down and technique may assist in the visualization of the vein and insertion of a catheter. Lymph contamination of blood samples is possible but less likely at this site.[1,13]

Brachial sinus

The brachial sinus requires the extension of a forelimb but is otherwise easily accessible. A needle is inserted perpendicular to the humerus immediately dorsal to the prominent biceps brachii tendon.[14] With slight negative pressure, the needle is advanced towards the radiohumeral joint until blood is observed.[14] Lymphatic contamination is possible and a second sample should be attempted if it is observed. In large tortoises like sulcata, a vein may be visible and palpable immediately proximal to the insertion site for brachial sinus blood collection. This vein may be amenable to intravenous catheter placement with a butterfly catheter but it should be removed prior to the animal awakening and withdrawing the leg.

Tail veins

Dorsal, lateral and ventral tail veins can be utilized in turtles. Their presence and accessibility will vary among species but the dorsal tail vein is the most commonly used.[1] Insert a needle in a cranial direction at an approximately 45° angle along the dorsal midline of the tail. With slight negative pressure on the syringe, advance until the needle hub fills with blood. If a vertebra is felt against the needle, withdraw and redirect slightly. Blood samples from this site are commonly contaminated with lymphatic fluid.[1,15]

Intraosseous catheterization

Sites for intraosseous catheterization in turtles include the femur, tibia and the bridge of the carapace.[1,4] Placement in the bridge is difficult and often results in communication with the coelomic cavity instead of the marrow space as intended.[1,4] At other sites turtles will damage the catheter when the leg is withdrawn into the shell thus IO catheters are typically limited to extremely moribund turtles or those under anesthesia.

Begin with flexing of the stifle joint and palpate the distal femur and the proximal tibia to determine the choice of site for catheterization. Choose either the distal femur or the proximal tibia based on ease of access around the shell and the shape of the bones. Heavily curved bones may add difficulty in feeding a catheter. Aseptically prepare the skin around the stifle. Lidocaine (2%) can be injected at the site of catheterization to prevent pain.[4] Place a spinal needle firmly against the proximal tibial crest, or at the distal end of the diaphysis of the femur; depending on which bone you have chosen. With gentle rotation and minimal pressure drill into the bone. The stylet of a spinal needle avoids catheter plugs but hypodermic needles can be used in smaller patients. With a loss of resistance the catheter is now in the medullary cavity and can be carefully fed further in to the bone. Confirmation of placement includes the presence of bone marrow, a catheter that moves synonymously with the bone as the leg is moved, audible scraping of the catheter inside the bone as it is gently rotated and orthogonal radiographs. Flush with heparinized saline. Place a small amount of antibiotic ointment on the insertion site to avoid infection and secure the catheter with tape wrapped around the leg both distally and proximally. A splint or additional bandaging can be used for additional protection as needed.

Amphibians

Ventral tail vein

Salamanders have similar venipuncture options to lizards and the approach to the ventral tail vein is the same. Many salamanders also perform tail autotomy so sedation or anesthesia should be considered.[1,6,10] Insert a small needle along the ventral midline of the tail in with a 45° to 90° angle in a craniodorsal direction.[1,6,10] If the tail vertebrae are hit with the needle and no flash of blood has been observed, pull back and redirect.

Ventral abdominal vein

The ventral abdominal vein is often visible and can be used in many amphibians.[1,6,10,16] If not visible a Doppler may help in locating the vessel.[6,16] Insert a needle at a shallow angle in a cranial direction along the ventral midline about midway between the sternum and pelvis.[10] Blood collection is slow, and lymphatic contamination common. This vessel may be large enough to attempt intravenous catheterization in some species.

Intra-cardiac

The heart can be located by visualizing the heartbeat, with a Doppler, or with transillumination.[1,6,10] Anesthesia is recommended for cardiocentesis to minimize risk of trauma to the heart or great vessels.[1,6,10,16,17] Insert a small needle along the ventrum towards the heart and advance into the ventricle. Minute negative pressure allows the syringe to fill with each heartbeat. Contamination with clear or cloudy yellow pericardial fluid is possible and the sample should be discarded if observed.[10]

Femoral vein

The femoral vein may be used to collect blood in anurans.[17,18] This vein may be visualized on the inside of the hind leg and blood collected with a small needle.[17] In very large frogs this vein may be large enough for intravenous catheterization.

Lingual vein

The lingual vein has been used as a readily available venipuncture site in anurans and it can be attempted in larger salamanders as well.[1,6,10,16,18] Open the mouth with a soft speculum and withdraw the tongue. Visualize the vessels located on the ventrum of the tongue or on the buccal floor and wipe clean. Puncture a vessel with a small needle and allow blood to seep from the wound.[1] Collect this blood with a heparinized capillary tube. Use the tongue or digital pressure until clot formation has built.[1]

References

1. Mader DR. *Reptile Medicine and Surgery.* St Louis, MO: Elsevier Saunders;2006.

2. Wellehan JFX, Lafortune M, Gunkel C. Coccygeal vascular catheterization in lizards and crocodilians. *J Herpetol Med Surg.* 2004;14(2):26-28.

3. Mosley CA. Intravascular access options in reptiles: what's reasonable. *Proc Annu Conf North Am Vet Comm.* 2006:1654-1655.

4. Briscoe JA, Syring R. Techniques for emergency airway and vascular access in special species. *Semin Avian Exot Pet Med.* 2004;13(3):118-131.

5. Mosley CA. Anesthesia and analgesia in reptiles. *Semin Avian Exot Pet Med.* 2005;243-262.

6. Campbell TW. *Exotic Animal Hematology and Cytology.* 4th ed. Ames, IA: Wiley-Blackwell;2015.

7. Mayer J, Knoll J, Wrubel K, et al. Characterizing the hematologic and plasma chemistry profiles of captive crested geckos (*Rhacodactylus ciliatus*). *J Herpetol Med Surg.* 2011;21:68-75.

8. Music MK, Strunk A. Reptile critical care and common emergencies. *Vet Clin North Am Exot Anim Pract.* 2016;19:591-612.

9. Joslin JO. Blood collection techniques in exotic small mammals. *J Exot Pet Med.* 2009;18:117-139.

10. Dyer SM, Cervasio EL. An overview of restraint and blood collection techniques in exotic pet practice. *Vet Clin North Am Exot Anim Pract.* 2008;11:423-443.

11. Brown C. Cardiac blood sample collection from snakes. *Lab Anim.* 2010;39:208.

12. Heatley JJ, Russell KE. Box turtle (*Terrapene* spp.) hematology. *J Exot Pet Med.* 2010;19:160-164.

13. Norton TM. Chelonian emergency and critical care. *Semin Avian Exot Pet Med.* 2005;14(2):106-130.

14. Mans C. Venipuncture techniques in chelonian species. *Lab Anim.* 2008;37:303.

15. Wilkinson R, McArthur S, Meyer J. Clinical pathology. *Med Surg Tortoise Turtle.* 2004:141-186.

16. Wright K. Amphibian hematology. In: Wright KM, Whitaker BR, eds. *Amphibian Medicine and Captive Husbandry.* Malabar, FL: Krieger Publishing;2001:129-146.

17. Allender MC, Fry MM. Amphibian hematology. *Vet Clin North Am Exot Anim Pract.* 2008;11:463-480.

18. Sykes JM, Greenacre CB. Techniques for drug delivery in reptiles and amphibians. *J Exot Pet Med.* 2006;15:210-217.

Reptilian and Amphibian Hematology and Cytology: What Just a Drop Can Tell You

Pia Bartolini, CVT, VTS (CP- Exotic Companion Animal)

Session #2150

Affiliation: From Terrestrial & Arboreal, LLC, Melrose, FL 32666, USA.

Hematology and cytology have critical benefits in reptilian and amphibian patients. With the quick turnaround time for in-house diagnostics a tailored treatment plan can be formatted prior to the client and patient leaving the clinic. Often times our ectothermic patients are extremely small and obtaining large sample quantities are not feasible. The beauty of hematology and cytology is one drop of blood, one drop of fluid, and/or one piece of tissue can give you rapid diagnostic results without immense cost to the clinic. There are some challenges with reptilian and amphibian clinical pathology such as intrinsic and extrinsic factors that will affect the sample. Quality sample collection and preparation are imperative for good diagnostics results and interpretation.

Critical and Emergent Care in Reptiles and Amphibians

Erica Mede, CVT

Session #2164

Affiliation: From Chicago Exotics Animal Hospital, 3757 West Dempster Street, Skokie, IL 60076, USA.

Veterinary technicians and nurses are integral to the success of a critical care case. In the past decade, there has been a push to appreciate, identify, and better maintain critical patients and emergent situations in herpetiles. Reptiles and amphibians often require critical care during hospitalization that is more advanced than what a veterinary technician or nurse has experience in. Critical care that is often underappreciated or overlooked during veterinary nursing includes: species consideration in treatment; multimodal analgesic approach; thermoregulatory behavior and support; supplemental enteral nutrition via esophagostomy tubes and gavage feedings; an understanding of diagnostics to utilize in emergency; fluid therapy considerations including fluid choice; fracture stabilization; and other advanced skills required to effectively treat a critical patient.

It's a Gut Thing: Gastrointestinal Anatomy, Physiology, and Nutrition of Ferrets and Rodents

Micah Kohles, DVM, MPA

Session #2165

Affiliation: From University of Nebraska-Lincoln, 1400 R St, Lincoln, NE 68588, USA, Oxbow Animal Health, 29012 Mill Rd, Murdock, NE 68407, USA, Flatwater Veterinary Group, 255 S 84th St Lincoln, NE 68510, USA.

Introduction

The gastrointestinal (GI) function of ferrets and other rodent species is complex and dependent on proper nutrition. Having an understanding of the GI anatomy and physiology of ECM species is key to properly educating owners, in addition to treating/diagnosing many common diseases as most we see in clinical practice are directly or indirectly related to nutrition. In this lecture we will focus on ferrets, rodents, hedgehogs and others. We will discuss their GI tracts and what diets are in the market including the good, bad and ugly.

Ferret

Domestic ferrets (*Mustela putorius furo*) are members of the Order Carnivora, Suborder Caniformia, Family Mustelidae, Subfamily Mustelinae. The scientific name *Mustela putorius furo* roughly translates to "mouse eating smelly little thief." There are 3 species of ferrets (also known as polecats in Europe and Asia): the European polecat (*mustela putorius*), the Steppe or Siberian polecat (*Mustela eversmanni*), and the black-footed ferret (*Mustela nigrans*). They are related to martens, fishers, otters, wolverines, long tailed weasels, least weasels, mink, badger and skunks. Ferrets are obligate carnivores, meaning they are designed to kill and eat whole prey items, which include all parts of the killed animal. They are specialized, with elongated bodies and short limbs, for the pursuit of warm blooded prey in holes and tight spaces. Ferrets tend to eat small meals throughout the day and will hide, cache, excess food in their dens or other hiding places.

Ferrets are obligate carnivores with a very short intestinal tract that lacks a cecum and ileocolic valve. Additionally, they are sometimes referred to as super carnivores, because of anatomic, physiologic and behavior traits adapted to a strict carnivore lifestyle. The gastrointestinal (GI) system of a ferret is essentially a long tube from the stomach to the rectum which allows food to pass too quickly for efficient absorption. The entire GI tract is under vagal and sacral innervation. The short length of the overall GI system of ferret's results in a rapid transit time of approximately 3-4 hours.

The gut flora is simple, and therefore GI upset or diarrhea secondary to antibiotic usage is uncommon. Like humans, ferrets are spontaneous secretors of hydrochloric acid which differs from many other predators such as dogs and cats. In addition, they spontaneously produce proteolytic enzymes and the GI tract remains active even under anesthesia. Ferrets do have the ability to vomit, but often do not even in the presence of obstruction or other pathology. Because of their rapid transit time and need for small frequent meals, a prolonged fasting, greater than 3-6 hours, is neither required nor recommended. It should be noted that ferrets with insulinomas may become hypoglycemic much quicker, even less than 3 hours, and therefore require special pre-operative care.

Ferrets have binocular vision, a very broad auditory range, and extremely sensitive olfactory specialization which make them excellent predators. They also have a hinge like temporomandibular joint, large mastication musculature and a large bite force to body mass ratio which allows them to predate larger sized prey.

Ferrets have typical carnivore dentition and oral anatomy with elongation of the carnassial blade, reduction or loss of post carnassial molars, and shortening of the rostrum in comparison to skull length. The upper incisors are slightly longer than the lower incisors when the mouth is closed. The canines are very pronounced, as in other carnivores. The lower premolars as well as first and second upper premolars all have two roots whereas the fourth upper premolars, the carnassial teeth, have three roots. The lower second molar is very small single rooted tooth. Dental formula: I (3/3) C (1/1) PM (3/3) M (1/2) = 34

Ferrets' teeth are precisely interdigitated, which is inline with the bladed shape of their teeth. Their teeth are designed to shear and tear animal tissue and are not designed for grinding. Previous research (Church) has pointed to commonly fed extruded diets predisposing ferrets to dental disease or exacerbating what could be consider normal dental wear. It is possible that the shape of their teeth and design for shearing are not in line with the chiseling and grinding process utilized to effectively masticate dry extruded kibble. Studies evaluating the differences between feral ferrets and domesticated ferrets have proven this difference.

The stomach of ferrets is simple and shaped similar to the dog. The stomach fits in the curve of the liver and has an identifiable pylorus. The two limbs of the pancreas are connected at the midline near the pylorus. It can expand to accommodate large amounts of food even though ferrets typically do not gorge themselves like some other carnivore species. The antrum of the stomach provides the mixing and propulsion of contents for gastric emptying.

The small intestine is quite short (~182-198 cm in length) and consists of three sections all of which contain goblet and villi cells. This short length is one factor that results in such a decreased GI transit time in ferrets. The major duodenal papilla is found ~3 cm from the pylorus and serves as the common opening for the bile and pancreatic ducts. The small intestine is made up of the duodenum, jejunum and ileum. Atropine does not block jejunal motility, as this movement is mediated by hormonal activity. Gastric acid secretion is only reduced by 30% in the presence of atropine. The ileum and jejunum are indistinguishable on gross examination and may be referred to as the jejunoileum, which ends at the ascending colon.

The entire large intestine of a ferret, which includes the colon and rectum, is approximately 10 cm in length. Ferrets lack both a cecum and ileocolic valve. This junction is inferred by the presence of the anastomoses of the jejunal artery with the ileocolic artery. The ascending, transverse, and descending colon make up the entire colon with the descending portion being the largest.

Nutrition

It is important to look at the evolutionary diet of ferrets to try and better understand their nutritional needs. Although there is no documented proof, it can be assumed that a ferret's diet would reflect that of the polecat. This would be composed of mostly small and medium sized mammals, but could also include a smaller portion of birds, carrion, reptiles, amphibians, eggs, crustaceans, mollusks, insects, worms and fish. Although some evidence reveals ferrets will feed on a small amount of seasonal plant based foods, their diet is obviously that of animal prey. They are, as we have discussed, strict obligate carnivores.

In view of the previous ferret GI discussion, it is abundantly clear how unique and inefficient their GI tract is and how important feeding an appropriate diet is to their health. To compensate for their inefficient GI system ferrets, require a very concentrated diet which is high in protein and fat as well as low in fiber and carbohydrates. The only carbohydrate sources in a ferret's natural diet are from the gut contents of their prey. Fat, when metabolized, releases twice as much energy as carbohydrate or protein metabolism and therefore should be the main source of calories in a ferret's diet.

While there has been some advancement in ferret diets, the majority of animals are still fed a dry extruded kibble. The majority of these diets contain higher than recommended levels of grains which provide starch and are required for kibble shape and stability as part of the extrusion process. This process consists of finely grinding ingredients and then mixing them into dough which is extruded through a tube under significant heat and pressure. The heating and processing of the animal proteins can potentially make them less digestible. While the author is not aware of peer-reviewed literature backing up such claims there has been ongoing discussion about the potential that dry extruded kibble that contain grains could be related to: 1) urolithiasis, due to diets high in plant proteins; 2) insulinomas, relating to the potential impact of carbohydrates on predilection for the formation this disease; and 3) excessive dental wear, since dry extruded kibble has been shown to cause an excessive amount of dental wear in relation to semi moist or natural diets.

Unfortunately, there are limited feeding options currently available to ferret owners. Although feeding a whole prey or raw diet is not owner-friendly, this is the best possible alternative for providing a complete natural diet for ferrets. It is important to educate owners about the potential risks of raw diets, such as bacterial (Salmonella, et.) and parasitic contamination which can be a zoonotic issue. There are a few semi moist diets on the market as well which, if they fit the nutritional profile, might be applicable but could predispose the ferret to periodontal disease if proper dental care is not maintained and most semi-moist diets still contain higher than recommended levels of carbohydrates.

Unlimited fresh, clean water is vital as ferrets require approximately three times as much water as dry matter. It is a good idea to offer ferrets water in more than one manner, such as a sipper bottle and bowl, to stimulate increased water intake. Some ferrets enjoy playing in their water so a heavy crock/bowl is recommended. Actual water consumption can be significantly affected by the diet provided (dry versus moist) as well as other environmental factors. Supplements or additives should not be added to ferrets' water.

Rat

The Norway or brown rat (*Rattus norvegicus*) belongs to Suborder Myomorpha (mouse-like rodents). Originating in temperate regions of central Asia, rats are nocturnal, burrowing, and colonial. Like most other myomorph rodents, rats are naturally omnivorous, eating grasses, seeds, grain, and occasionally invertebrates. Rats are quiet, clean, and easy to maintain. They make excellent pets, especially if socialized from an early age.

Rats feed primarily at night and sleep during the day. They tend to avoid new foods therefore diet changes should be made gradually. Diet should be uniform commercial pellet/extruded kibble (rodent block) supplemented with moderate amounts of fresh fruits and vegetables and infrequent offerings of seeds and insects. Seed-based diets should be avoided to prevent selective feeding, malnutrition, and obesity. Protein recommendations vary from 16% to 20%. Rats practice coprophagy, which provides nutrients such as B-vitamins and bacterial protein. Young rats are reared communally and nursing responsibilities are shared. Weaning occurs at 21 days of age.

Like other rodents, rats are monophydont (have only one set of teeth). Their dental formula is 2(I 1/1, C0/0, P 0/0, M3/3). The ratio of upper incisor to lower incisor length is approximately 1:3, which inexperienced practitioners often mistake for overgrowth of the lower incisors. Food is grasped in the forepaws, chipped by the incisors, and ground down by the molars. Only one side of the mouth is in apposition at a time during mastication. Cheek tissue can be drawn into the diastema, closing off the back of the mouth to allow gnawing without debris passing into the oral cavity.

The esophagus enters the stomach in the middle of the lesser curvature. The stomach is monogastric and divided into two parts by a limiting ridge. Situated at this junction, the limiting ridge prevents a rat from vomiting. Myomorphs such as the rat are unable to vomit also, in part, due to the crural sling (muscle) and the innervation of the diaphragm; therefore, pre-operative fasting is not required. The non-glandular forestomach is lined by squamous epithelium, while the distal glandular part is lined by fundic glands containing chief and parietal cells. Like other myomorphs, rats have a relatively large cecum and long colon. The cecum is not septate as in other species. GITT for the rat is 12-24 hours.

Mouse

The domesticated mouse (*Mus musculus*) originated in temperate Asia. Mice belong to the suborder Myomorpha (mouse-like rodents) and are primarily nocturnal. They are social and prefer to live in small groups dominated by a single adult male. Mice are omnivorous in the wild, eating grasses, seed, grain and occasionally invertebrates. Mice naturally hoard food items; thus food consumption is hard to measure. A commercial rodent chow (rodent block/lab block) is recommended over a seed-based diet. This may be supplemented with fresh foods and limited seeds/insects as treats and to encourage foraging. The diet should contain 4-5% fat and 16% or more protein for maintenance; 7-11% fat and 20-24% protein for breeding. Rodent chow can be supplemented with fresh foods in moderation. Fresh water should be available at all times (monitor water bottles for malfunction). Like all rodents, mice practice coprophagy, which provides nutrients produced via hindgut fermentation.

Mice have elodont incisors and anelodont check teeth. The mouse dental formula is 2(I 1/1, C0/0, P 0/0, M3/3), and the upper/lower incisor length ratio is 1:3. Mice are monogastric, but there is a division (the limiting ridge) between the forestomach and the glandular stomach. The forestomach is lined by squamous epithelium and the glandular stomach by cardiac, fundic, and pyloric glands. Like other myomorphs, mice have a relatively large cecum and long colon. Because of the lower esophageal sphincter, the limiting ridge, crural sling, and innervation of the diaphragm, mice cannot vomit. The liver of the mouse is composed of four lobes: median (divided into left and right portions), left and right lobes, and caudate lobe. The life span of mice averages 1.5-2 years; up to 3 years is reported.

Gerbil

The Mongolian gerbil (*Meriones unguiculatus*) is a Myomorph rodent native to the desert regions of Mongolia and northeastern China. It is a social animal that is active during the day and night. Gerbils are adapted to a desert environment, require little water and produce only small amount of concentrated urine. Their feces are small, dry, and hard. For this reason, gerbils are nearly odorless and require less cleaning than do other rodents.

Wild gerbils have a variable, seasonal diet of leaves, seeds, and insects. If captive gerbils are fed seeds they tend to selectively feed and therefore consume an incomplete diet. A pelleted/extruded rodent chow with 16-22% protein is recommended. For reproduction, dietary protein may need to exceed 20%. Although wild gerbils can obtain almost all the water they need from available fruit and vegetable matter, commercial pellets do not contain adequate moisture to meet these needs. The captive diet should be supplemented with small amounts of fresh fruit and vegetables such as apples and carrots as well as small amounts of seeds and insects. Gerbils should have multiple sources of fresh, clean water available always. Fat in the diet should be 4% or less, as gerbils are prone to high cholesterol levels. Gerbils appear to have a higher need for magnesium than rats, and develop alopecia and increased susceptibility to seizures when fed an Mg-deficient diet. Gerbils are not coprophagic when fed a nutritionally complete diet.

Hamster

Hamsters are Myomorph rodents of Middle Eastern and Asian origin. Several different species of hamster are kept as pets: the Syrian or golden hamster (*Mesocricetus auratus*); the closely-related dwarf hamsters-the Russian dwarf, Siberian or Djungarian hamster (*Phodopus sungorus*), the Campbell's Russian dwarf hamster (*Phodopus campbelli*), and the Roborovskii (Robo) or desert hamster (*Phodopus roborovskii*); and the Chinese or striped hamster (*Cricetulus griseus*). Hamsters are generally solitary and nocturnal, although dwarf hamsters may live in family groups.

Hamsters are nocturnal, burrowing animals that prefer a deep layer of bedding. The word hamster comes from the German word hamstern, which means to hoard. Hamsters naturally hoard food therefore the cage must be cleaned frequently to remove hoarded food and prevent ingestion of moldy food items. Hamsters are desert adapted; however, they are not heat tolerant; they burrow to escape heat. They will hibernate at low environmental temperatures, which may be mistaken for extreme debilitation or death. Hamsters live 1.5-3 years depending on species and environmental factors.

Hamsters are omnivorous and coprophagic. The feeding pattern of adults is typically a 5-min feeding period, followed by a 2-hr fast. Diet in the wild includes grains, seeds, plants, and some invertebrates. Dietary protein requirements are lower for hamsters than for mice/rats. Pet hamsters are usually fed a commercial hamster mix containing 16-20% protein, 60-65% carbohydrate, and 5-7% fat, which can be supplemented with fruit, vegetables, nuts, and animal protein (egg, cheese, insects). Seed- based diets are common, but lend towards selective feeding and unbalanced nutrition. Commercial pellets are less likely to result in nutritional deficiency, obesity, and diabetes. Hay (oat is a good choice) is offered to increase fiber consumption, and sugary treats are kept to a minimum. Hamsters should have wood, branches, etc. to gnaw on for enrichment, and fresh water must be available at all times (water bottle malfunction is relatively common).

The hamster dental formula is 2(I1/1, C0/0, P0/0, M3/3). The four prominent incisors are elodont while the cheek teeth are anelodont. The crown of the mandibular incisors is longer than the crown of the maxillary incisors, commonly mistaken to be overgrown. The yellow color of the incisors is due to iron pigments. The cusps of cheek teeth retain food and thus hamsters are prone to dental caries formation (males > females).

Hamsters have large cheek pouches, invaginations of the oral mucosa used for transporting food, bedding material, and young. When swimming, hamsters can inflate their cheek pouches in order to increase buoyancy. Cheek pouches extend to the shoulders/scapulae and are retracted by the buccinator muscle, which evolved from the trapezius muscle. Hamsters empty their cheek pouches by pushing cheek contents forwards with the paws. When weighing a hamster, the weight of food in the cheek pouches must be taken into consideration.

Hamsters have a distinct non-glandular forestomach (*pars cardiaca*) that resembles the rumen. Like a rumen, it has high pH and contains microorganisms for fermentation. The glandular stomach (*pars pylorica*), is like that of a monogastric animal. The differential in pH between the two compartments indicates that food does not mix between them. The presence of VFAs (mainly acetic acid) indicates that some fermentation does go on here, although food only stays in the stomach for about 1 hour. The forestomach is lined by squamous epithelium while the glandular stomach region is lined by glandular mucus membranes. Stomach bacteria are mainly Gram-positive with some Gram-negative coliforms. The two chambers are separated by a muscular sphincter which may regulate passage of food between the two sections. The distal esophagus has a muscular sphincter and enters the stomach just cranial to this constriction. The lesser curvature of the stomach is greatly reduced such that the cardia and the pylorus are very close; therefore, vomiting is impossible for hamsters.

The cecum is sacculated, containing bacterial microflora including *Bacteroides* spp. and *Lactobacillus* spp. plus a few coliforms and clostridia. The pH in the cecum is higher than in the non-glandular stomach, suggesting more active fermentation. The liver has four lobes: right and left medial, and right and left lateral. Hamsters are permissive hibernators; low environmental temperatures stimulate food gathering, and at 41°F and below they enter deep sleep.

Degu

The degu (*Octodon degus*) is native to the western foothills of the Andes in northern and central Chile. It is the most populous mammal in Chile, where it is considered an agricultural pest. Degus are rodents belonging to the Suborder Hystricognathi (porcupine-like rodents), based upon jaw musculature and skull structure. They belong to the Family Octodontidae, and Genus Octodon, which refers to the figure-eight shape of their cheek teeth. Its other names include brush-tailed rat and trumpet-tailed rat. Degus are semi-fossorial, diurnal, and do not hibernate.

Degus are adapted anatomically and behaviorally to a fibrous, low-carbohydrate diet. The wild diet includes a wide variety of plants, grasses, roots, seeds, fruit, and livestock droppings. Captive diet should consist of high quality hay, limited high-fiber herbivore rodent pellets (eg, chinchilla pellets), a small amount (20% of daily intake) of greens such as kale, collard, romaine lettuce and dandelion greens, broccoli, green beans, and roots such as beets, carrots, parsnips, sweet potatoes, and turnips. Degus produce the enzyme L-gulonolactone oxidase and therefore do not require dietary supplementation with vitamin C.

Improper diet is a major cause of disease in captive degus. To reduce the risk of diabetes and cataracts, foods containing sugar (apples, raisins, other fruits, breakfast cereal, honey treat sticks, etc.) should be avoided. Starchy foods may also pose a risk.

Degus are similar anatomically and to both guinea pigs and chinchillas. They possess large, powerful incisors which are yellow to orange from iron pigmentation. The dental formula is 2(I 1/1, C 0/0, P 1/1, M 3/3). Both the incisors and the cheek teeth are elodont (aradicular).

The degu stomach is simple and glandular, and the intestine is the primary site of nutrient absorption. Degus possess a large, functioning, haustrated cecum where microbial fermentation of ingesta occurs. The colon lacks haustra, and colonic absorption of water from digesta results in very dry feces. GITT is very fast at only 5.1 hours. Degus are cecotrophic and coprophagic: one study showed that 38% of feces produced in 24 hr period were reingested; 87% of this activity occurred at night. Cecotrophy and coprophagy begin as early as 3 days of age.

Prairie Dog

The black-tailed prairie dog (*Cynomys ludovicianus*) is a stout, short-legged ground squirrel native to the Great Plains of the US. Prairie dogs belong to Suborder Sciuromorpha (squirrel-like) rodents. They are herbivorous, having a functioning cecum. In their natural habitat, prairie dogs eat primarily grasses, prairie herbs and weeds, and occasionally roots and insects. Prairie dogs are diurnal and spend much of the day foraging for food; they do not store food in their burrows.

Captive prairie dogs should receive unlimited grass hay (Diversity is beneficial) and a measured quantity of pelleted food; prairie dog specific is preferred but are often hard to source, but high quality rabbit pellets are also suitable. Alfalfa, fresh greens, and vegetables can be offered as treats. Calorie consumption for the prairie dog varies throughout the year, increasing in the fall; caloric intake should be regulated to prevent obesity. Fresh water should always be available, provided daily in both a bowl or sipper bottle. Fluid consumption varies depending on diet. Wild prairie dogs can get most of the water they need from the plants they eat; however, dry pelleted diets necessitate more water intake than fresh foods do.

Prairie dogs are in the squirrel family. Their incisors are elodont (aradicular) whereas their cheek teeth are anelodont. The dental formula for the prairie dog is 2(I 1/1, C 0/0, PM 2/1, M 3/3). Incisor enamel is normally orange-yellow in color. Prairie dogs are hindgut fermenters that require adequate dietary roughage. They are also coprophagic. Normal droppings resemble those of guinea pigs, dry and oval in shape. Prairie dogs have a triad of scent glands surrounding the anus. Their white papillae resemble fly larvae when they are winked. These are not removed as an elective procedure in prairie dogs. Prairie dogs are more difficult to handle and care for than other pet rodents. As a native species, they are also illegal to keep in some states.

True hibernation does not occur in prairie dogs, but dormancy may occur for brief periods during cold weather. Dormancy can also be stimulated in captivity by sudden drops in temperature. Prairie dog pups appear above ground at approximately 5-6 weeks, and are weaned at about 5-7 weeks. The majority of prairie dog pups in the pet trade are harvested each spring from the wild. Prairie dogs don't breed well in captivity.

Hedgehog

The African pygmy hedgehog (*Atelerix albiventris*) is native to West and Central Africa, one of many hedgehog species found in Africa, Europe, Asia, and the Middle East. Hedgehogs are monogastric omnivores, consuming a variety of invertebrates, frogs, lizards, snakes, mice, eggs, fruit, and fungi. They are generally solitary and nocturnal. They inhabit a wide range of habitats including grassland, scrub, savannah, and suburban gardens. They live in brush piles, rock crevices, and burrows. They are able to climb, dig and swim. The African pygmy hedgehog can hibernate during periods of cool, dry weather, remaining torpid for periods of up to six weeks but emerging during warm intervals.

Hedgehogs exhibit a unique behavior known as self-anointing or anting, which can be triggered by any new or strange object or substance. The new object is licked until thick, frothy saliva accumulates. The saliva is then carefully placed onto the spines of both sides of the body with the tongue. This elaborate process may continue for up to 20 minutes. The behavior may serve as a sexual signal (confined to the breeding season in the wild) or protection against predators. Substances that cause self-anointing behavior are typically novel or irritating (eg, hedgehogs have been observed to rub toads against their spines). This behavior has been observed in hedgehogs as young as 15 days of age.

Hedgehogs have 36 teeth. The dental formula is: 3/2 incisors, 1/1 canines, 3/2 premolars, and 3/3 molars. Hedgehogs have a simple stomach and no cecum. The gastrointestinal transit time averages 12-16 hours. African pygmy hedgehogs do not appear to have a physiologic need to hibernate and will not do so if kept warm and well fed. They may enter hibernation if the environmental falls below 18°C (65°F). Metabolism, heart rate and respiratory rate will decrease, while synthesis of endogenous heparin increases, preventing thrombus formation. This decrease in metabolism renders hedgehogs more susceptible to infection, therefore, pet hedgehogs should be prevented from hibernating.

The natural diet of hedgehogs includes insects, worms, snails, slugs and occasionally small vertebrates and fruit. There is limited knowledge of the nutritional requirements of hedgehogs. They can be successfully maintained on high quality dry low-fat cat food. Commercial diets are also available and should be supplemented with earthworms, mealworms, crickets, and a small amount of chopped fruits and vegetables. Hard-boiled or scrambled eggs, evaporated milk, cottage cheese, wax worms, pinky mice, and vegetable or meat baby foods provide additional variety to the diet. An insect or meat-only diet should be avoided in order to prevent metabolic bone disease.

Hedgehogs can learn to use a sipper-type watering system or water can be provided in a shallow pan. Provide fresh water from multiple sources daily. Offer food in the evening and is usually limited to the amount an adult will consume by morning. Use heavy crocks and food bowls that cannot be tipped. Ad lib feeding, over-consumption, and inactivity commonly lead to obesity in African hedgehogs. The problem can become so severe that the patient can no longer fully roll up. Obesity can lead to other metabolic derangements and health problems, including hepatic lipidosis. Treat obesity by gradually reducing the quantity of food until the amount offered is consumed overnight. Limit foods high in fat: meal worms, wax worms, etc. Offer live insects in the cage to promote normal hunting and burrowing activity.

It's a Fiber Thing: Gastrointestinal Anatomy, Physiology, and Nutrition of Exotic Companion Mammal Herbivores (Rabbits, Guinea Pigs, and Chinchillas)

Micah Kohles, DVM, MPA

Session #2166

Affiliation: From University of Nebraska-Lincoln, 1400 R St, Lincoln, NE 68588, USA, Oxbow Animal Health, 29012 Mill Rd, Murdock, NE 68407, USA, Flatwater Veterinary Group, 255 S 84th St Lincoln, NE 68510, USA.

Abstract: The anatomy and gastrointestinal physiology of rabbits, guinea pigs, and chinchillas are different than other exotic companion mammals. While these species have similar gastrointestinal function and nutritional needs, there are some unique species traits.

Introduction

The anatomy and gastrointestinal physiology of rabbits, guinea pigs, and chinchillas are different than other exotic companion mammals. These species are strict herbivores, classified as hindgut (cecum and colon) fermenters, concentrate selectors, and are designed to ingest large amounts of high-fiber food. As prey species, maintaining a small body size is advantageous. Their unique gastrointestinal system, with its rapid transit time and ability to differentiate particulates of fiber, allows them to remain small and active, while surviving on high-fiber, low-energy density plant materials. While these species have similar gastrointestinal function and nutritional needs, there are some unique species traits. Gastrointestinal disease in these species is a very common presenting issue in clinical practice. It often is associated with dysbiosis and can present as both primary and secondary disease.

Rabbit

Domestic rabbits (*Oryctolagus cuniculus*) are herbivores, concentrate selectors, and are classified as hindgut (cecum) fermenters. They are mostly crepuscular and nocturnal feeders. They are anatomically and physiologically adapted to handle significant amounts of low-energy density fibrous food and to effectively utilize the nutrients found in a high-fiber diet; however, due to their small body size, they are unable to store large amounts of food material and therefore utilize a unique process to rapidly eliminate fiber for the gastrointestinal system. Their gastrointestinal system allows for the intake of large amounts of low-energy density, high-fiber food that is separated based on particle size. The easily fermentable component of the diet is retained, while the slowly fermentable components of the diet (preeminently cellulose based plant fiber) are rapidly eliminated. This colonic separation mechanism (CSM) allows for the production of two types of stool: a dry, hard, high-fiber fecal pellet and an enlarged, soft fecal pellet that is covered in mucous and referred to as a cecotroph (eg, cecal, night stool). In rabbits, it is commonly referred to as the wash back CSM. Rabbits, like all herbivores, have a symbiotic microbe relationship with a diversity of gut flora (primarily anaerobic), as they produce limited mammalian enzymes to break down the cellulose components of their plant based diets. Some of the most common issues with domestic rabbits, both in companion and commercial animals, are related to the gastrointestinal system. The symbiotic relationship rabbits have with the micro flora in the cecum allows them to effectively ferment plant structural-carbohydrates which provide their main energy source.

Rabbits have a large cornea and laterally placed eyes, which give them a panoramic field of view when grazing and foraging. Food is selected based on smell and tactile perception of the lips, as their visual field does not include the area just under their nose.[1] As concentrate selectors, rabbits will choose the most tender, succulent plant parts which are nutrient dense and lowest in fiber to meet their nutritional needs, owing to a high metabolic rate. Rabbits have hypsodont or open-rooted (constantly growing) incisor and cheek, including premolar and molar, teeth.

Rabbits' incisors are designed to apprehend and cut through vegetation. The large upper incisors have two smaller incisors behind them (peg teeth). The occlusion of the upper and lower incisors allows the teeth to be constantly sharpened. The premolars and molar teeth, also known as cheek teeth, are designed to process hard, fibrous foods. The fibrous food is ground between the cheek teeth with jaw movements up to 120 per minute.[1] This repetitive, rapid chewing movement helps to maintains proper dental health. It should be noted that when rabbits ingest cecotrophs, normal mastication does not occur and the cecotrophs are swallowed intact.[2] When a rabbit is fed an unsuitable diet, such as low-fiber muesli mixes and minimal or no hay, improper dental wear can quickly occur, leading to dental disease. A suitable diet of high-fiber hay and high-fiber, uniform pellets will help control the growth of the teeth and prevent overgrowth and malocclusion. As concentrate selectors, rabbits, like deer, have a predisposition to select the lower fiber, more tender portions of their diets which, if they are fed cereal based or mixed pellets, can lead to selective feeding and malnutrition.

If low-fiber mixed cereal or seed based foods are eaten, the rapid chewing activity is no longer required, which can result in dental disease, such as malocclusion and root elongation, as well as gastrointestinal issues. As the teeth elongate, secondary to inadequate wear, the jaws are forced apart and the radiographic appearance of the diastema increases. Over time, the jaws begin to no longer occlude properly and mandibular movement is reduced. The condition deteriorates as the teeth become mobile and maloccluded, resulting in the development of spurs that may lacerate and ulcerate the tongue and buccal cavity.

The stomach is J-shaped and large compared to most monogastrics, comprises roughly 15% of the GI tract, and has both a well-developed cardiac sphincter that prevents vomiting and a muscular pyloric area.[1,3] The pH of the stomach is low and ranges from 1 to 6. A variety of factors affects the pH, including the site of determination, the presence or absence of cecotrophs, the duration of time from feed intake to measurement, and the age of the rabbit.[4] The pH of weanling rabbits is higher (5-6.5), which permits the passage of bacteria through the stomach to the hindgut to colonize the cecum.[1]

The small intestine, made up of the duodenum, jejunum (longest), and ileum, is short compared to other species at ~3 meters in length and comprises ~12% of the total gastrointestinal tract.[5,6] Transit time through the jejunum and ileum is 10-20 and 30-60 minutes respectively.[4] The duodenum is the largest of three segments but shortest in length (~3-5 inches in rabbits).[7] The distal end of the ileum is thickened and referred to as the sacculus rotundus, which marks the intersection of the ileum, colon, and cecum. This location signals the beginning of the large intestine and links to the ampulla coli, which forms an intersection of the distal ileum, cecum, and proximal colon. The sacculus rotundus is unique to the rabbit.[1] Located at the junction of the ileum and the large intestine is the ileocecal valve, which regulates the movement of material from the small intestine into the large intestine and prevents the retrograde movement of intestinal contents.

The small intestine is the site where the greater part of digestion and absorption takes place by passive and active transportation throughout the mucosa.[4] The stomach and small intestine in rabbit's digest and absorb nutrients very similarly to monogastric animals with hydrochloric acid and pepsin as initiators of digestion. Pancreatic enzyme production, such as amylase, is minimal, with most coming from saliva or bacterial production within the cecotroph.[1] A key function of this component of a rabbit's GI system is the digestion of cecotrophs. Cecotrophs contain essential nutrients, including volatile fatty acids, hydration, protein, vitamins and amino acids. They are coated in a protective mucous which protects them from the acidity of the stomach prior to absorption in the small intestine. The transit time of the small intestine is rapid compared to other herbivorous species and fiber is quickly moved to the cecum and large intestine.

The rabbit cecum is very large and has a capacity roughly ten times that of the stomach, making up around 40% of the total gastrointestinal tract.[3] It is a thin-walled organ that folds on itself multiple times and has an internal surface made up of a long spiral fold (sometime referred to as spiral valve) that is continued into the beginning of the colon, an area known as the *ampulla coli*. The distal tip of the cecum is known as the vermiform appendix. This tube-like area is narrow and thickened compared to the cecum, and contains significant amounts of lymphoid aggregates which secrete bicarbonate ions into the cecum. The bicarbonate ions act as a buffering agent for the volatile fatty acids formed during cecal fermentation.[2]

The colon is divided into three sections, the ascending, transverse, and descending colon. Because the *fusus coli* (located anatomically at what is referred to as the transverse colon) forms such a natural division between the morphologically and functionally distinct ascending and descending sections of the rabbit colon, many physiology texts have abandoned the 3-section description and simply use the terms proximal and distal colon.[1,8] The colon is a very functional component of the hindgut and is characterized by sacculations (haustra) and bands (taeniae). The ascending colon has four sections that contain small protrusions, approximately 0.5 millimeters in diameter that are commonly referred to as warzens. These wart-like protrusions are believed to be unique to rabbits. They potentially represent an increase in the surface area of the colon that would favor absorption and also may assist in mechanical separation of gut contents.[1] The *fusus coli* has a mucosa that is 4-5 times thicker than the descending colon and contains ganglion aggregates.[1] The *fusus coli* is commonly referred to as the pacemaker of the hind gut and controls the CSM with its retrograde and normo-grade peristaltic activities that occur during the formation of soft and hard feces. The descending colon mucosa is smooth, and while it contains numerous goblet cells, does not contain obvious specialization, and usually contains hard fecal pellets.[1]

Rabbits do not completely ferment fiber. They instead utilize the wash back CSM to sort out indigestible fiber (based on size) and expel it from the body. The rhythmic contractions of the cecum, constantly mix the ingesta as it undergoes fermentation by a large diversity of bacterial species, predominantly gram-negative anaerobic *Bacteriodes* spp and gram-positive anaerobic *Clostridiales*. The effects of bacterial fermentation include the production of volatile fatty acids (VFA) including acetic, formic, propionic, and butyric acids. These are a major energy source for rabbits. A portion of these VFAs are contained in cecotrophs and utilized upon re-ingestion, while another portion is absorbed directly across the cecal mucosa.

There is an almost continuous flux of material between the cecum and the proximal colon. The CSM mixes fluid and nutrients through vigorous peristalsis in both normo and retrograde directions. These contractions result in large indigestible fibrous particles accumulating in the center of the colon where they are rapidly transported along the colon to the rectum for defecation, usually within 4 hours of ingestion. Conversely, smaller, more fermentable fibrous particles accumulate at the periphery of the colon where through retrograde peristaltic contractions of the haustra, they are returned to the cecum for further fermentation.

Periodically there is a dramatic change in the peristaltic activity of the cecum and colon as the retrograde movement of smaller fiber particles ceases and a large quantity of the cecal content is expelled into the colon. These contents are excreted, usually once to twice daily, as soft, mucous covered cecotrophs. This process is controlled by the *fusus coli*. The ability of the colon, via the colonic separation mechanism, to rapidly eliminate large indigestible fiber particles and retain smaller, more digestible fiber for further fermentation, makes the rabbit an extremely efficient herbivore, capable of surviving on very low-quality forage.

The cecotroph fermentation process depends heavily on an appropriate diet and the action of resident bacteria and protozoa which are vital to the gastrointestinal health of the rabbit. The diversity of bacterial flora and other organisms within the cecum is substantial and includes many bacterial species dependent on location within the cecum lumen or wall, unidentified anaerobic species, ciliated and flagellate protozoa, and a rabbit specific yeast.[9] There is still much to learn and understand about the microbial populations of the rabbit hindgut but recent research has shown the hindgut microbial environment to be heavily anaerobic bacterial species and that *lactobacillus* are not normal flora for rabbits. Fermentation produces volatile fatty acids, vitamins B, C, and K, and proteins which are eventually digested in the small intestine after cecotrophy, which occurs once or twice a day, usually at night.

In healthy and physically able rabbits, the large, soft mucous covered cecotrophs are ingested directly from the rectum (midnight snack). Multiple factors, including stimulation of rectal mechanoreceptors, perception of the cecotroph odor, and the blood concentrations of various metabolites and hormones, all stimulate the rabbit to directly ingest the cecotrophs. Higher fiber diets increase cecophagy; whereas, high-protein and low-fiber diets decrease cecophagy. Once eaten, the acidity of the stomach breaks down the protective mucous covering of the cecotroph. Continued digestion occurs in the small intestine where cecotrophs represent an important component of the diet providing protein, water and vitamins.

Guinea Pig

Guinea pigs (*Cavia porcellus*) are obligate herbivores and have a large tongue and narrow oral cavity. Dental disease is a common malady, as discussed with rabbits, and proper diet (specifically, ample fiber intake) is key to decreasing the likelihood of dental issues. Guinea pigs, like rabbits, are hind gut fermenters and are coprophagic. While the belief is that coprophagy is an important function in guinea pigs, its contribution to overall nutritional needs has not been fully characterized.[10] Coprophagy in rabbits is a key source of B vitamins and a means of optimizing the availability of proteins.[2] Guinea pigs though are not cecotrophic, and unlike rabbits, are not reliant on the key amount of B vitamins found in cecotrophs. Rabbits require a dietary source of 3 of the 10 B vitamins; whereas, guinea pigs require a dietary source of 7 of the 10.[11] Small herbivores utilize 1 of 2 primary mechanisms for the colonic separation mechanism necessary for coprophagy. This capability permits rapid passage of less digestible food particles while retaining microorganisms and more digestible food particles in the cecum, to allow adequate opportunity for fermentation.[12] Rabbits, as previously discussed, utilize a multi-directional peristaltic motion to separate components of fiber based on size and mix these with other ingesta, bacteria, and water to concentrate more fermentable particles and eliminate larger particles of fiber (wash back CSM). Guinea pigs and chinchillas, which are more efficient are fermenting fiber than rabbits, utilize a CSM process sometimes referred to as mucous trap CSM, in which bacteria from the cecum are trapped in the mucus of the colon with few to no food particles and returned to the colon through antiperistalsis.[10] The mucous trap process is less efficient than the mechanisms utilized by rabbits.[13] Guinea pigs and chinchillas have a colonic furrow in the ascending colon, in which the concentration of bacteria and nitrogen is twice as high as in the lumen.

The stomach of guinea pigs contains the same four regions as found in rabbits, the cardia, fundus, body, and pylorus. The small intestine of guinea pigs is made up of 3 sections (duodenum, jejunum, and ileum), as found in most species. They are visually non-distinguishable, with the jejunum being the longest section (~90-100 cm). The cecum is the largest component of the guinea pigs' gastrointestinal tract and contains upwards of 65% of total contents.[14] The bacterial flora is similar to the rabbit, being primarily gram-positive with anaerobic lactobacilli being the primary organisms in the large intestine.[2] As in the rabbit, a large number of additional organisms including yeast, protozoa, and other bacterial species are prevalent. Gastric emptying time is ~2 hours and total gastrointestinal transit time is ~20 hours (8-30 hours), and considerably longer (up to 60 hours) if coprophagy is considered.[15,16]

Chinchilla

Chinchillas are obligate herbivores, hind gut fermenters, and cannot vomit like rabbits and guinea pigs. They are more similar to the guinea pig than the rabbit in terms of gastrointestinal anatomy and physiology. Their dentition is also more similar to guinea pigs than rabbits, as they have a larger grinding surface owing to a naturally higher fiber and coarser diet than rabbits. As compared to the guinea pig, the jejunum and descending colon are long.[10] The colon is highly sacculated and the cecum large, yet the cecum holds less contents of the overall large intestine in chinchillas compared to rabbits and guinea pigs.[10,17]

Nutrition

The discussion of the gastrointestinal system of herbivorous exotic companion mammals above makes it very clear how integral proper diet is to the health of all small herbivores. The keys to a proper small herbivore diet are the following:

- High-quality, high-fiber grass hay (~70-75%).
- Grass hay should be fed free choice and available at all times. Any high-quality grass hay such as timothy, orchard, oat or organic meadow hay is appropriate and feeding a diversity species is optimal from both a nutritional and enrichment perspective.
- These species will also use hay for bedding and nesting material. It is important to differentiate these different uses and ensure the animals always have access to clean fresh hay for ingestion.
- High-fiber uniform pellet (~20-25%).
- Timothy, mixed grass or alfalfa based depending on life stage. Always feed only the recommended amount.
- It is important to avoid products that contain added seeds, fruits, or vegetables, or that utilize additional ingredients (typically starch or carbohydrate based) required for certain manufacturing techniques.
- Fresh or dried greens (~10-15%).
- Only appropriate species should be fed and all should be carefully cleaned. Organic products are preferred. They focus should be on greens and only limited and infrequent amounts of vegetables and even less fruit. A good resource of appropriate greens is the House Rabbit Societies website. www.rabbit.org
- Treats (~0-5%).
- Low-sugar, high-fiber treats. Avoid artificially colored, flavored or preserved products.
- Unlimited fresh, clean water.
- Multiple sources, such as water bottle and crock, should be available at all times.

High-quality, high-fiber grass hay should be available at all times. Fiber is essential to maintain proper gut health, and to stimulate gut motility (insoluble fiber only).[18,19] Although there are multiple grass hay varieties available on the market, there is minimal nutritional difference between them, and therefore, different species, such as timothy, orchard, oat, and mixed are appropriate. Hay is a product of nature, and therefore, there will be variances in color, texture, feel, and taste. Because of this, it is important to transition between hay bags and offer different varieties of hay. Hay not only provides a key component of the nutritional needs of rabbits and other small herbivores, but it also encourages natural behaviors such as foraging and grazing. Stimulating these natural behaviors may diminish boredom based behaviors, increase activity, and also provide a sense of security to the animal.

There are many diets available to pet owners, the two most common being seed/cereal based muesli mixes and hay based, high-fiber uniform pellets. These species are all, to some degree, concentrate selectors, and if offered a mix, will selectively feed on certain items from the mix, which will lead to improper and unbalanced nutrition. These mixes are often high in carbohydrates and simple sugars, which can lead to obesity and GI issues, as well as low in fiber, which slows gastrointestinal motility and decreases normal teeth wear. Uniform hay-based fortified pellets, high in fiber and low in protein, are recommended, as they eliminate selective feeding, thereby guaranteeing a proper vitamin and mineral intake.

Fresh and dried greens are an important and enjoyed component of a rabbit's and small herbivore's diet. They provide an excellent source of water, enrichment and other plant based nutrients. Varying types and textures is a great enrichment tool; however, there should be some transition period when introducing new types of greens. It is recommended to completely wash and clean all greens, and organic products are preferable.

Although treats are not nutritionally required for rabbits or other small herbivores, limited appropriate treats can be beneficial for strengthening the pet/owner bond, training, and enrichment. It is important to limit the number of treats and ensure they are low in protein, fat, calcium, and sugar. Some examples of inappropriate treats are nuts, seeds, popcorn, bread, crackers, and yogurt drops. High-fiber grass and alfalfa based treats are appropriate in limited amounts, as are small amounts of dried fruits/veggies. There are far more inappropriate treats on the market, but good products do exist.

Unlimited fresh, clean water is vital and the water requirements of rabbits have been estimated at 50-100 ml/kg/day dependent on their diet.[20] Fresh, clean water should be available to small herbivores at all times. It is a good idea to offer herbivores water in more than one manner, such as a sipper bottle and bowl/crock, to stimulate increased water intake. Water should be changed daily, as bottles/bowls/crocks will dirty with normal drinking. Actual water consumption can be significantly affected by diet and other environmental factors.

References

1. Harcourt-Brown F. *Textbook of Rabbit Medicine.* Oxford, UK: Butterworth-Heinemann;2002.

2. Cheeke P. *Rabbit Feeding and Nutrition.* Orlando, FL: Academic Press;1987.

3. Jenkins J. Feeding recommendations for the house rabbit. *Vet Clin North Am Exot Anim Pract.* 1999;2(1): 143-151.

4. Carabano R, Piquer J, Menoyo D, Badiola I. The digestive system of the rabbit. In: de Blas C, Wiseman J, eds. *Nutrition of the Rabbit.* 2nd ed. Oxfordshire, UK: CABI;2010.

5. Alworth L, Harvey S. Chinchilla anatomy, physiology and behavior. In: Suckow M, Stevens K, Wilson R, eds. *The Laboratory Rabbit, Guinea Pig, Hamster and Other Rodents.* San Diego, CA: Academic Press;2012:955-966.

6. Sohn J, Cuoto M. Rabbit anatomy, physiology and behavior. In: Suckow M, Stevens K, Wilson R, eds. *The Laboratory Rabbit, Guinea Pig, Hamster and Other Rodents.* San Diego, CA: Academic Press;195-215.

7. Wingerd B. *Rabbit Anatomy and Dissection Guide.* Eden Prairie, MN: Blue Door Publishing;2007:57-67.

8. Snipes T, Clauss W, Weber A, Hornicke H. Structural and functional differences in various divisions of the rabbit colon. *Cell Tissue Res.* 1982;225:331-346.

9. Davies R, Davies J. Rabbit gastrointestinal physiology. *Vet Clin North Am Exot Anim Pract.* 2003;6:139-153.

10. Quesenberry KE, Carpenter JW, eds. *Ferret, Rabbits, and Rodents Medicine and Surgery.* 3rd ed. St. Louis, MO: Elsevier Saunders;2012.

11. Stevens C, Hume I. Contributions of microbes in vertebrate gastrointestinal tract to production and conservation of nutrients. *Physiol Rev.* 1998;78:393-427.

12. Bjornhad G, Snipes R. Colonic separation mechanism in lagomorph and rodent species – a comparison. *Zoosystem Evol.* 1999;75:275-281.

13. Franz R, Kreuzer M, Hummel J, et al. Intake, selection, digesta retention, digestion and gut fill of two coprophageous species, rabbits (*Orytolagus cuniculus*) and guinea pigs (*Cavia porcellus*), on a hay only diet. *J Anim Physiol Anim Nutr (Berl).* 2010;95(5):564-570.

14. Hargaden M, Singer L. Guinea pig anatomy, physiology and behavior. In: Suckow M, Stevens K, Wilson R, eds. *The Laboratory Rabbit, Guinea Pig, Hamster and Other Rodents.* San Diego, CA: Academic Press;2012:575-602.

15. Harkness J, Murray K, Wagner J. Biology and diseases of guinea pigs. In: Fox J, Anderson L, Loew, Quimby F, eds. *Laboratory Animal Medicine.* 2nd ed. San Diego, CA: Academic Press;2002:203-246.

16. Jilge B. The gastrointestinal transit time in the guinea pig. *Z Verdsuchstierk.* 1980;22:204-210.

17. Wolf P, Schroder A, Wenger A, Kamphues J. Nutrition of the chinchilla as a companion animal-data, influences and dependences. *J Anim Physiol.* 2003;87:129-133.

18. Irlbeck N. How to feed the rabbit (*Oryctolagus cuninulus*) gastrointestinal tract. *J Anim Sci.* 2001;79: 343-436.

19. Griffiths M, Davies D. The role of the soft pellets in the production of lactic acid in the rabbit stomach. *J Nutr.* 1963;80:171-80.

20. Brewer N, Cruise L. Physiology. In: Manning P, Ringer D, Newcomer C, eds. *Biology of the Laboratory Rabbit.* 2nd ed. San Diego, CA: Academic Press;1994:63-70.

Section 2

Plenary Session

Nicole Johnson, DVM and
Micah Kohles, DVM, MPA
Session Chairs

Sponsored By:

Developing Moment By Moment Resilience Using the Principles of Improv

Jeff Thoren, DVM, PCC, BCC

Session #2167

Affiliation: From Gifted Leaders, LLC, Chandler, AZ 85248, USA.

Abstract: How can the principles of improv, help you become more adaptable and resilient? According to author Robert Poynton, "Improvisation is what life does. Nothing living, from a bacterium to a blue whale, has a script for their life. Somehow or other, every living being copes with untold complexity without a plan, and always has." In this interactive session you will experience how you can have order without control (just like nature does) by applying three simple, interconnected principles.

Manuscript reprinted with permission: From Thoren J. Developing Moment by Moment Resilience Using the Principles of Improv. *Proc Annu Conf North Am Vet Comm.* 2017.

Introduction

im • pro • vise: verb. *To fabricate out of what is conveniently on hand.*

Consider for a moment that every day of your life is a piece of improv on the stage of life. In his book, *Do Improvise*, author Robert Poynton shares the following introduction to the promise of improvisation as an important life skill:

The unexpected is our constant companion. Blind alleys, unheralded turns and serendipitous connections are everyday stuff. However hard we work to avoid it, we are constantly accosted by things we didn't plan for, from a puncture to a deflating economy. Life is a whirling torrent for which nobody has a script.

Such a complex world demands an improvised response. The good news is that we are, in fact, naturally good at creative adaptation and there is a substantial measure of it in all that we do.

The 'offer' here, as an improviser would say, is to learn to understand, trust and develop our ability to improvise. To become both more willing and more adept at flexing, adapting and adjusting to what we have, rather than wishing we had something else.

Improvisers have developed the ability to create a coherent story, incorporating interruptions, objections and curve balls seamlessly as they go. They are able, constantly, to build ideas and relationships, even in the most demanding circumstances.

This represents quite a shift in attitude. Instead of trying to bend events (and people) to your will, you focus on discovering a way to work with whatever you find. This can be very liberating. You realize that you don't have to know everything. That you don't always have to be on top of things, that you can allow yourself to be 'in the thick' of them. That a simple shift from 'yes, but' to 'yes, and' can make you far more creative. At heart this attitude is founded on humility and acceptance, which helps you become more compassionate, especially to yourself. It is a lighter way of being.

People think of improvisation as unruly and chaotic but, in fact, it is a discipline with a system of rules that guide your conduct. This improvisational discipline has a number of uncommon virtues. It assumes that events around you, and the people that provoke most of those events, are beyond your control. As a result it discourages you from wasting any time or energy trying, in vain, to control them. Instead, it focuses on how to use whatever they give you, even if it is a challenge or an objection, and trains you to see these as opportunities. It pays attention to what you can control, which is your own attitude and response to whatever happens.

Principles of Improv

The principles of improvisation are very simple. They are not rocket science but, instead, more like making a paper airplane ... with a little instruction everyone can do it.

The principles can be summed up in six words. They are:

- Notice More
- Let Go
- Use Everything

With presence of mind, you can apply these ideas at any moment, in any context.

If left to its own devices, our human mind habitually wanders away from the present moment. When we're not in the here and now, we dwell in the past, grasping and replaying it, or we project into the future, trying to anticipate the unknown (and often catastrophizing).These habitual thought patterns don't serve our ultimate well-being.

Getting caught up in self-referential thinking can get in our own way. Applying the three principles of improv can help us get out of our own way and into the flow of life. When we can get out of our own way, we're happier and more engaged with the world, we're more compassionate, and as a result we can perform at our best.

The principles of improv help us to be more mindful. Mindfulness is defined as "paying attention in a particular way: on purpose, in the present moment, and non-judgmentally." When we can consciously stop regretting things from the past or worrying about what we're going to do in the future and "get out of our own heads," we can be fully present in the moment. This allows our mind to return to an ideal state where it is both relaxed and alert at the same time. When this happens, three wonderful qualities of mind naturally emerge: calmness, clarity, and happiness.

Notice More

> *The aim of life is to live, and to live means to be aware, joyously, drunkenly, serenely, divinely aware.*
>
> – Henry Miller

We can choose what we pay attention to. This is about deepening your capacity to be present ... in the world and to the people around you.

Effective approaches to notice more include:

- Get out of your head and "lean into your senses" (eg, seeing, hearing). The goal here is to consciously let more in vs. filtering stuff out.

- Tap into your intuition and notice more about your own body (eg, what are you experiencing at a heart level or at a gut level?). As an example, we usually experience emotions more vividly in the body than in the mind. Bringing attention to the body enables a high-resolution perception of our emotions. The better we can perceive our emotions, the better we can manage them.

In their book, *Improvisation at the Speed of Life*, TJ Jagodowski and David Pasquesi (TJ & Dave) share the following insight about the art of improv:

"It's not that tough. You just have to take the next step on the path in front of you. You don't have to invent. You just have to notice. You don't have to play the whole show. You just have to play this moment. Now this moment. Now this moment."

Noticing more helps you cultivate greater awareness, find new possibilities, and see opportunities you never expected.

Let Go

> *Life is about not knowing, having to change, taking the moment and making the best of it without knowing what's going to happen next. Delicious ambiguity.*
>
> – Gilda Radner

We need to let go of baggage from the past and extrapolations into the future, because they stop us paying proper attention to the present, and the present is where we live.

How might you (and other people around you) benefit if you were to ...?

- Let go of judgment and leaping to conclusions. Instantly evaluating what is happening through the lens of the past gives you no chance of adapting creatively.

- Let go of shadow stories (not the actual story, but our misinterpretation of it) and assumptions. Projecting shadow stories and incorrect assumptions into the future prevents us from paying attention to what is actually happening.

Many of us operate from a conscious or subconscious belief that we need to be in control. We would likely think that we could never do improv because there is no script. The reality is that we likely get ourselves in more trouble when we try to script or control things then when we let go.

TJ & Dave have this to say about why we try to control things ...

"The reason we want to pre-plan the scene is fear. We're afraid of not knowing what it is about or that it won't be funny. Maybe the audience won't like us or our scene partner will get more laughs. In order to improvise well, we need to remember that those reactions to fear are not helpful."

They go on to offer this piece of encouragement ...

"Rest easy. We need to embrace this whole endeavor as a process of improving, learning, and not knowing it all. Improvisation is something you can enjoy wildly while you're doing it and, at the same time, know 'Hey, I'm not very good at this yet.' (It's like sex in this way.) The goal is to commit fully and trust in our partners. A good idea before taking the stage is to remind oneself and all involved that the goal is not to be funny or to be clever. The goal is to be present."

To what degree is fear keeping you from letting go?

Letting go means giving up the need to control outcomes and going where the journey takes you. Not having a script can actually be quite liberating and we all have the potential to be naturally good at it.

Use Everything

Be where you are; otherwise you will miss your life.
– Buddha

Everything is an offer.

Whatever you face, regardless of how inviting or irksome it may seem, is usable in some way. This denies you any wiggle room and forces you to look harder for what's useful.

To apply this principle, consider that …

- If you view your mistakes as offers you learn fast. You can see a mistake as a "mis-take," like an actor on a film set, and look at it as another attempt in an iterative process where you are striving to get better. If mistakes are opportunities, you don't need to make apologies, look for scapegoats or find excuses: you just get on with working out how to use them.

- Other people are constant sources of offers. They always bring interpretations, experience and perspectives that you don't have. If someone sees things differently from you, instead of thinking they are wrong, see it as an offer. Ask yourself how their point of view can enrich yours.

If you want to use what you have, an obvious and easy place to start is simply to ask yourself, "What have I got?" You might be surprised at how much you have overlooked.

Unforeseen consequences, accidents, disasters and delays call all be used to your advantage, if you are prepared to consider them as offers. Thinking in terms of using everything reminds you to let go of judgment.

Putting It All Together

Martin de Maat noted that "In order to improvise in front of an audience, you have to be accepting, involved in the moment, and courageous." Doesn't this also clearly speak to what's needed to be successful as a human being?

If every day of your life is indeed a piece of improv on the stage of life, by applying the three simple, interconnected principles outlined above, you can be adaptable and resilient without the need to be in control (which is an unrealistic expectation anyway). And you will also experience more authentic happiness, be more compassionate towards yourself and others, and be able to make the greatest possible contribution in your work by reaching your full potential.

All you have to do is remember to incorporate these three principles into your life:

- Notice More
- Let Go
- Use Everything

Improv, Mindfulness, and Emotional Intelligence

> *Between stimulus and response, there is a space. In that space lies our freedom and our power to choose our response. In our response lies our growth and our happiness.*
>
> – Viktor Frankl

The principles of improvisation, mindfulness, and emotional intelligence are interrelated. As mentioned earlier, the principles of improv help us to be more mindful. But it's also true that the practice of mindfulness facilitates our ability to notice more, let go, and use everything.

In his book, *Search Inside Yourself*, author Chade-Meng Tan observes the following connections between mindfulness and emotional intelligence.

Self-awareness is the key domain of emotional intelligence that enables all the others. Self-awareness depends on being able to see ourselves objectively, and that requires the ability to examine our thoughts and emotions from a third-person perspective, not getting swept up in the emotion, not identifying with it, but just seeing it clearly and objectively. This requires a stable and clear, non-judging attention.

The way to train this quality of attention is something known as "mindfulness meditation." Mindfulness is defined by Jon Kabat-Zinn as "paying attention in a particular way; on purpose, in the present moment, and non-judgmentally." The calmness and clarity that results from mindfulness meditation increases the "space" that Viktor Frankl refers to in the quote above.

Cultivating emotional intelligence begins with mindfulness. We use mindfulness to train a quality of attention that is strong both in clarity and stability. We then direct this power-charged attention to the physiological aspects of emotion so we can perceive emotion with high vividness and resolution. The ability to perceive emotional experience at a high level of clarity and resolution builds the foundation for emotional intelligence. And we live happily ever after.

Chade-Meng Tan suggests two ways to experience a taste of mindfulness:

1. Simply bring gentle and consistent attention to your breath for two minutes. Pay attention to the process of breathing and every time your attention wanders away, just bring it back very gently.

2. Sit without an agenda for two minutes. Shift from "doing" to "being," whatever that means to you, for just two minutes. Just be.

References

1. Wolkin J. 4 Lessons in Mindfulness from the Olympians. How phrases like "good sportsmanship" and "getting your head in the game" are really calls for being more mindful. The Mindful.org Blog. Available at: https://www.mindful.org/ Accessed June 30, 2017.

2. Poynton R. *Do Improvise. Less Push. More Pause. Better Results. A New Approach to Work (and Life).* London, UK: The Do Book Company;2013.

3. Jagodowski TJ, Pasquesi D, Victor P. *Improvisation at the Speed of Life: The TJ and Dave Book.* Chicago, IL: Solo Roma Inc;2015.

4. Tan C-M. *Search Inside Yourself. The Unexpected Path to Achieving Success, Happiness (and World Peace).* New York, NY: HarperCollins Publishers;2012.

5. Brewer J. You're Already Awesome. Just Get Out of Your Own Way! TEDx Rock Creek Park. Available at: https://www.youtube.com/watch?v=jE1j5Om7g0U Accessed June 30, 2017.

Development of an Antibiotic Policy in a Zoological Medicine Service and Approach to Antibiotic Dosing Using Minimum Inhibitory Concentration Data

Stephen J. Divers, BVetMed, DZooMed, DipECZM (Herpetology), DipECZM (Zoo Health Management), Dipl ACZM, FRCVS,
Izidora Sladakovic, BVSc (Hons I), MVS,
Jörg Mayer, DVM, MS, DipECZM (Exotic Mammals), Dipl ACZM, DABVP (Exotic Companion Mammal Practice),
Susan Sanchez, BSc, MSc, PhD, FRSB

Session #2101

Affiliation: From The Department of Small Animal Medicine and Surgery (Divers, Sladakovic, Mayer) and Athens Veterinary Diagnostic Laboratory (Sanchez), College of Veterinary Medicine, University of Georgia, Athens, GA 30602, USA.

Abstract: Antimicrobial resistance is a major veterinary concern, and yet despite widespread literature and policies advocating appropriate antibiotic stewardship in domesticated veterinary medicine, there appears to have been very little progress in zoological medicine, especially companion exotic practice. An example of an antibiotic policy for a zoological medicine service is described, along with general advice on appropriate antibiotic use, including minimum inhibitory concentration (MIC)-determined drug dosing.

Introduction

Antimicrobial resistance (AMR) is a global veterinary and public health challenge, which has accelerated by the overuse of antibiotics worldwide. Increased antimicrobial resistance is the cause of severe infections, complications, longer hospital stays and increased mortality. Overprescribing of antibiotics is associated with an increased risk of adverse effects, more frequent re-attendance and increased medicalization of self-limiting conditions. Antibiotic overprescribing is a particular problem in primary care, as the vast majority of all antibiotic prescriptions are issued by general practitioners.[1]

In 2015, the American and European Colleges of Veterinary Internal Medicine (ACVIM, ECVIM) published a consensus statement on therapeutic antimicrobial use and AMR.[2] This consensus provided the veterinary profession with up-to-date information on the pathophysiology, diagnosis and treatment of clinically important animal diseases. The information was derived from evidence-based medicine with specialist commentary, prior to peer review, and subsequent publication in the Journal of Veterinary Internal Medicine.

"The epidemic of antimicrobial resistant infections continues to challenge, compromising animal care, complicating food animal production, and posing zoonotic disease risks. While the overall role of therapeutic antimicrobial use in animals in the development of AMR in animal and human pathogens is poorly defined, veterinarians must consider the impacts of antimicrobial use in animals and take steps to optimize antimicrobial use, so as to maximize the health benefits to animals while minimizing the likelihood of AMR and other adverse effects."[2]

Despite recommendations and policies from many veterinary associations (including the American Veterinary Medical Association, British Veterinary Association, European Commission, etc), and many taxa-based veterinary organizations (encompassing companion dogs, cats, and horses, as well as production poultry, cattle and swine), it seems that zoological medicine, and especially companion exotic animal care, has largely failed to produce appropriate policies. A review of the Association of Avian Veterinarians (AAV), Association of Exotic Mammal Veterinarians (AEMV), and the Association of Reptilian and Amphibian Veterinarians (ARAV) websites failed to demonstrate any policy on antibiotic stewardship. The author hopes that this presentation will energize the zoological companion animal specialty and associations to exhibit greater leadership in this area.

Review of cases referred to the Zoological Medicine Service, Veterinary Teaching Hospital, University of Georgia (ZMS) has indicated that virtually all animals seen by a previous veterinarian were treated with antibiotics. The vast majority of antibiotic drugs were dispensed without attempting to identify whether any bacterial disease was actually present (by Gram stain, culture, PCR, cytology, or histopathology). Furthermore, many cases were treated with multiple antibiotic drugs and/or courses of therapy using advanced antimicrobials (eg, fluoroquinolones, 3rd or 4th generation cephalosporins). Our subsequent case investigations suggest that the majority of antibiotic prescriptions were unnecessary (no bacterial disease present) or inappropriate (inappropriate choice of drug, drug resistance, inappropriate dose and/or duration).

The reasons for such poor therapeutic decisions may stem from client pressure, lack of knowledge, expertise, or confidence on the part of the veterinarian, and/or the client pressure to prescribe something, anything! Whatever the reasons, the reliance on broad spectrum antimicrobials implies a low level of skill and expertise on the part of the veterinarian. Consequently, changes in policy and culture are required, and we exotic animal veterinarians appear to lag well behind our domestic animal peers in this regard, who have already advocated for, and in many cases adopted, appropriate policies with positive results.[3,4] One such policy has been developed and instigated in the ZMS (Table 1), as part of a larger hospital policy on antibiotic use.

Table 1. University of Georgia Zoological Medicine Service Antibiotic Policy.

Antibiotics are to be prescribed by a veterinarian for two possible reasons, (I) prevention of infection and (II) treatment of bacterial disease

Prevention of Infection
Antibiotics should only be dispensed to prevent infections IF the animal is (a) severely compromised (eg, neutrophil/heterophil count <1, or some other severe tissue compromise/exposure, eg, severe burns or other trauma), OR (b) receiving short term IV or IO antibiotics intraoperatively to prevent infection (eg, orthopedics, enterotomy). Tier 1 drugs can be used.

Treatment of Infection
Some attempt to identify the bacterial cause (Gram stain, cytology, culture and sensitivity, MIC PCR) is required prior to prescribing any antibiotic for therapeutic purposes. Tier 1, first-line dr can only be used initially while pending culture and sensitivity or MIC results. Tier 2 drugs can only be used with a specific indication by culture and sensitivity or MIC testing that demonstrat poor efficacy or resistance to tier 1 drugs. Tier 3 drugs should not be used unless specifically authorized following the existing hospital protocol on "Restricted Use Antimicrobial Agents foı Multidrug Resistant Infections."

Table 1 Continued.

Tier 1 Antibiotic Drugs (First line options)
Uses (a) prevention of infection in severely compromised animals, (b) intraoperative IO or IV administration to prevent infection (eg, orthopedics), and (c) first line therapeutic antimicrobial treating infections, while awaiting culture and sensitivity or MIC results;
Trimethoprim, sulfonamides or combinations
Tetracyclines (eg oxytetracycline, doxycycline)
Basic/potentiated penicillins (eg ampicillin, penicillin, amoxicillin/clauv)
Metronidazole
Lincosamides (eg, lincomycin, clindamycin)
Aminoglycosides
1st and 2nd generation cephalosporins (incl intraop cefazolin)
1st generation quinolones (eg, oxolinic acid, nalidixic acid)

Tier 2 Antibiotic Drugs (Only if sensitivity/MIC results indicate resistance to tier 1)
Only to be used if culture and sensitivity or MIC testing indicate tier 1 drugs are ineffective;
3rd generation cephalosporins (ceftazidime, ceftiofur)
Penicillinase-resistant penicillins (eg, methicillin)
Advanced penicillins (eg, piperacillin, cerbenicillin, ticarcillin)
2nd generation fluoroquinolones (eg, enrofloxacin, orbifloxacin, ciprofloxacin)
Florfenicol, Chloramphenicol

Tier 3 Antibiotic Drugs (RESTRICTED)
Only to be used in cases of multidrug resistance where specified hospital criteria have been met authorization has been granted for their use.
Glycopeptides (eg, vancomycin)
Carbapenems (eg, imipenem)
Oxazolidonones (eg, linezolid)
4th generation and above cephalosporins (eg, defepine)
Ketolides (eg, telithromycin)
Lipopeptides (eg, daptomycin)
Ansamycins (eg, geldanamycin)
3rd generation and above fluoroquinolones (eg, levofloxacin)

Prevention of Bacterial Infection

Antibiotics can be used prophylactically to prevent infection; however, such use can increase the chances of commensals developing resistance. There is no argument in favor of prophylactic antibiotics because of poor hospital facilities, equipment, patient preparation or surgical technique. Routine sterilizations (castrates and spays) performed appropriately do not warrant antibiotics. Even in cases of corrective surgery (including enterotomy and orthopedics) where there are concerns regarding the development of post-operative infection, short-term pre- or peri-operative antibiotics (usually intravenous) are more effective and less likely to result in resistance, compared to several days of post-operative treatment.[5,6] There appears to be little to no evidence to support the routine use of post-operative antibiotics to prevent infection following clean-contaminated surgeries.

Broad spectrum antibiotic therapy has frequently been recommended in cases of severe integumental (eg, burns) or gastrointestinal (eg, intestinal breach and gut-origin sepsis) compromise; however, a systematic review and meta-analysis in human burn victims concluded that antibiotic prophylaxis is not currently recommended for patients with severe burns other than perioperatively.[7] In addition, strict criteria and recommendations exist for antibiotic use in cases of gut-origin sepsis.[8] Antibiotics are also prescribed in cases of life threatening immunosuppression, typically associated with radiation or chemotherapy. Antibacterial prophylaxis is only generally recommended for patients expected to have neutrophil counts of $< 1 \times 10^9/L$ for > 7 days, unless other factors increase risks for complications or mortality.[9]

Therapeutic Treatment of Bacterial Disease

In order to successfully treat a bacterial disease several facets must be confirmed:

1. Does a bacterial disease actually exist?
2. Can a safe and effective antibiotic drug dose be delivered?
3. Can therapy be continued to the point of cure?

If the answer to any of these questions is no, then antibiotic therapy is unlikely to be successful and should not be pursued.

Confirmation of bacterial disease

Any oral swab from a snake is likely to reveal *Pseudomonas*, but that merely indicates the presence of bacteria, not that stomatitis is present. This is especially critical when dealing with commensal organisms. To demonstrate bacterial disease, it is generally necessary to demonstrate a host pathologic response (ie, cytology, histopathology) as well as the bacterial cause (ie, Gram, culture, PCR).

Antimicrobials should not be prescribed until some attempt has been made to identify a bacterial disease. In most cases this entails cytology and culture (but PCR may be used concurrently especially if antibiotics have been recently given and bacteria are no longer viable for culture). Such laboratory confirmation can take up to several days, and therefore in-house Gram stains should be routinely employed to help direct initial antibiotic drug selection. Indeed, in those cases where clients baulk at the costs of professional cytology and microbiologic culture, in-house Gram stain and cytology can be performed at lower costs to external labs; however, confirming the presence of a bacterial disease should be considered an integral, critical requirement prior to prescribing antibiotics (just like analgesia is considered an integral, non-optional component of anesthesia and surgery).

Initial drug selection and dose

Antibiotic therapy should be delayed until the animal is normothermic, and hydrated. If disease has been confirmed, then initial Gram stains, site of infection and likely organism involved should be used to help direct initial drug selection from a list of first-tier drugs (see Tables 1 and 2). For example, gram-positive infections should be treated using a basic penicillin, 1st and 2nd generation cephalosporin, or a potentiated sulphonamide. The opinion that the more modern drugs (eg, carbenicillin, ceftazidime) are more efficacious against susceptible bacteria than older products is simply untrue. If the patient is severely compromised then bacteriostatic drugs may be less favored (eg, tetracyclines, macrolides, lincosamides) over bactericidal options. Aminoglycosides and fluoroquinolones are not indicated and typically less effective against gram-positives. Conversely, first-tier drugs against gram-negative infections would include potentiated sulphonamides or aminoglycosides, while metronidazole is indicated for suspect anaerobes. More advanced or modern drugs (Table 1 tier 2) should not be used as first line options, and their use is best restricted to cases where culture and sensitivity dictate their necessity. Furthermore, those drugs that play a critical role in the treatment of multidrug resistant diseases should be restricted and their use not permitted without permission from hospital authorities (Table 1, tier 3).

Regarding dose, many published formularies are available, but a majority of doses are anecdotal and not pharmacokinetically derived; however, where such data exists in the species of interest (or a closely-related species) it is often the most reliable source for dose determination. A quick search through Google Scholar can often be most rewarding for indicating pharmacokinetic research. Extrapolation from one species or taxa to another, or allometric scaling may be unavoidable but the increased risks of inappropriate dosing has to considered.

Table 2. Standard bacterial sensitivities to various classes of antibiotics.

Penicillins	
Penicillin G	Bactericidal, inhibit mucopeptide synthesis in the cell wall. Active against most spirochetes and gram-positive and gram-negative aerobic cocci, but not penicillinase producing strains. Effective against some aerobic and anaerobic gram-positive bacilli such as *Bacillus anthracis*, *Clostridium* sp. (not *C. difficile*), *Listeria*, *Fusobacterium*, *Actinomyces*, *Corynebacterium*. Inactive against most gram-negative aerobic and anaerobic bacilli, mycobacteria, and mycoplasmas.
Ampicillin Amoxicillin	Increased activity against many strains of gram-negative aerobes not covered by penicillin G, including some *E. coli*, *Klebsiella*, *Salmonella*, *Shigella*, and *Haemophilus*. Some anaerobic activity against Clostridia. Generally ineffective against *Pseudomonas aeruginosa*, *Serratia*, indole-positive *Proteus* (*Proteus mirabilis* is susceptible), *Enterobacter*, *Citrobacter*, *Acinetobacter*, *Mycobacterium*, and *Mycoplasma*. Clavulanate (B-lactamase inhibitors) improves activity against B-lactamase staphylococci, *H. influenzae*, *Neisseria*, *Moraxella catarrhalis*, *Bacteroides*, and *Klebsiella*.
Cloxacillin Methicillin	The penicillinase-resistant penicillins are active against staphylococci and streptococci (including pneumococci) but not against enterococci. Their primary use is for infections caused by penicillinase-producing staphylococci.
Carbenicillin Piperacillin Ticarcillin	Broad-spectrum, with similar activity to ampicillin but are also active against *Enterobacter*, *Serratia*, and *P. aeruginosa*. Piperacillin is also active against many strains of *Klebsiella*. The addition of clavulanate to ticarcillin and of tazobactam to piperacillin adds activity against *Klebsiella*, *Serratia*, and *Bacteroides* and against B-lactamase-producing strains of staphylococci, *H. influenzae*, and *N. gonorrhoeae*.
Cephalosporins	
1st generation cephalothin cefazolin cephapirin cephradine cephalexin cefadroxil	Bactericidal, inhibit mucopeptide synthesis in the cell wall. Excellent coverage against most gram-positives, *Streptococcus* (except *Enterococcus*), *Staphylococcus*, *Proteus mirabilis* and some *E. coli*, *Klebsiella*, *Actinobacillus*, *Pasturella*, *Haemophilus*, *Shigella* and *Salmonella*. Most anaerobes are very susceptible, except *Bacteroides fragilis*. Most species of *Corynebacteria* are susceptible. Group D streptococci/enterococci (*S. faecalis*, *S. faecium*), Methicillin-resistant *Staphylococcus*, indole-positive *Proteus* sp., *Pseudomonas* sp., *Enterobacter* sp., *Serratia* sp. and *Citrobacter* sp. are generally resistant.
2nd generation cefaclor cefamandole cefonicid ceforanide cefuroxime	Similar gram-positive coverage as 1st generation, but also good activity against *Bacteroides* fragilis. Gram-negative spectrum variable.

Table 2 Continued.

3rd generation cefotaxime moxalactam cefoperazone ceftizoxime ceftazidime ceftriaxone ceftiofur cefixime	Similar gram-positive activity of 1st and 2nd generations, plus expanded, but variable, Gram-negative activity. Ceftazidime and cefoperazone are active against most strains of *Pseudomonas aeruginosa*.
Aminoglycosides	
Amikacin Gentamicin Kanamycin Neomycin Streptomycin Tobramycin	Bactericidal, bind to the 30S ribosome and inhibit bacterial protein synthesis. Only effective against aerobic gram-negative bacilli and staphylococci. Activity against streptococci and anaerobes is poor. Aminoglycosides may be used in combination with a penicillin in staphylococcal and streptococcal infections. Streptomycin, neomycin, and kanamycin lack activity against *Pseudomonas aeruginosa*, whereas gentamicin, tobramycin, and amikacin have good activity. An aminoglycoside should always be added to a B-lactam antibiotic when treating serious *Pseudomonas* infections. Amikacin has the same spectrum of activity as gentamicin and tobramycin but is less susceptible to enzymatic inactivation. Therefore, amikacin may be useful in treating gentacin or tobramycin resistant infections.
Macrolides and lincosamides	
Azithromycin Clarithromycin Clindamycin Erythromycin Lincomycin Tilmicosin	Primarily bacteriostatic and bind to the 50S subunit of the ribosome and inhibit bacterial protein synthesis. Active against aerobic and anaerobic gram-positive cocci, and against gram-negative anaerobes. Erythromycin is active against most gram-positive cocci (including anaerobes), but many human *Staphylococcus aureus* strains are now resistant. Erythromycin is also active against *Mycoplasma*, *Chlamydophila*, *Bordetella*, *Corynebacterium*, *Campylobacter*, and *Treponema*. Although it has activity against anaerobic gram-negative bacilli, its activity is much less than that of clindamycin. Clarithromycin and azithromycin have enhanced activity against *Haemophilus influenzae* and activity against *Mycobacterium avium*-intracellulare.Clindamycin has poor activity against *Mycoplasma* but much greater activity against anaerobic bacteria, especially *Bacteroides* sp. CNS penetration into the brain and CSF is poor.
Tetracyclines	
Doxycycline Oxytetracycline Tetracycline	Bacteriostatic, bind to 30S subunit of ribosome and inhibit bacterial protein synthesis. Effective against many B-hemolytic streptococci, nonhemolytic streptococci, gram-negative bacilli, rickettsiae, spirochetes, *Mycoplasma*, and *Chlamydophila*. Bacterial resistance to one tetracycline indicates likely resistance to the others. Used primarily for urinary tract infections, rickettsial, chlamydial, *Mycoplasma*, *Shigella*, *Brucella*, and *Vibrio* infections. Considered an alternative to penicillin treatment for syphilis in humans and maybe useful for rabbits.
Fluoroquinolones	
Ciprofloxacin Enrofloxacin	Bactericidal and inhibit the activity of DNA gyrase. Activity against Enterobacteriaceae and staphylococci, *P. aeruginosa*, *Mycoplasma*, *Chlamydia*, and some streptococci, but not reliably active against anaerobes. Resistance has been noted, particularly with *P. aeruginosa* and methicillin-resistant *Staphylococcus aureus*. Resistance to one fluoroquinolone generally means resistance to all.
Potentiated sulphonamides	
Trimethoprim-sulfa combinations	Bacteriostatic, both drugs block the folic acid metabolism cycle of bacteria and are synergistic together. Sulfonamides are competitive inhibitors of the incorporation of p-aminobenzoic acid. TMP prevents reduction of dihydrofolate to tetrahydrofolate. Trimethoprim-sulfa is active against most gram-positive and gram-negative organisms but is inactive against anaerobes. *Pseudomonas aeruginosa* is usually resistant.

Table 2 Continued.

Polypeptides	
Polymyxin B	The polypeptides are bactericidal antibiotics with activity against gram-negative aerobic bacilli including *Pseudomonas aeruginosa*. Polymyxin B is not active against *Proteus* sp. or gram-positive organisms.
Bacitracin	Bacitracin is a bactericidal, inhibits cell wall synthesis. Active only against gram-positive organisms and very few gram-negatives. Used topically and may be effective orally in the treatment of *Clostridium difficile*.
Miscellaneous	
Chloramphenicol Florfenicol Thiamphenicol	Primarily bacteriostatic, binds to the 50S subunit of the ribosome and inhibits bacterial protein synthesis. Wide spectrum of activity against gram-positive and gram-negative cocci and bacilli (including anaerobes), *Rickettsia*, *Mycoplasma*, and *Chlamydophila*. Primary uses in serious *Salmonella* infections, *Haemophilus*, and *Pasteurella*, but also effective against *Mycoplasma*. Ineffective in meningitis caused by *E. coli* and other Enterobacteriaceae.
Metronidazole	Bactericidal, disrupts bacterial DNA and nucleic acid synthesis. Activity against most obligate anaerobes including *Bacteroides* sp. (including *B. fragilis*), *Fusobacterium*, *Veillonella*, *Clostridium* sp., *Peptococcus*, and *Peptostreptococcus*. *Actinomyces* is frequently resistant to metronidazole.
Rifampin	Bactericidal or bacteriostatic dependent upon bacterial susceptibility and drug concentration, inhibits DNA-dependent RNA polymerase. *Mycobacterium*, *Staphylococcus aureus*, *Neisseria*, *Haemophilus*, and *Rhodococcus*.

Modification of drug selection

Disc sensitivity testing provides a subjective in vitro assessment of antibacterial sensitivity (sensitive, intermediate, resistant), and although it provides a useful guide there are often times when in vivo results do not concur with disc sensitivity. Minimum inhibitory concentration (MIC) is much more reliable and provides a more objective, quantifiable sensitivity. MIC data are useful, not only for determining which drug should be selected from a group of efficacious options, but also for calculating the patient's required dose from pharmacokinetic data (Vd, Cmax).

Clinical application of MIC data using therapeutic factor

As a rule of thumb, clinical response correlates best for:

1. systemic infections if the attainable peak serum/plasma level is 2-4X MIC;
2. urinary tract infections, if attainable bladder level is 10X MIC (μg/ml);
3. biofilm in lumens 1000–1500X MIC.

In order to use MIC data to direct patient therapy by calculating a preferred drug dose (PDC), the following must be known, approximated, or calculated:

1. The cultured bacteria's MIC to specific antibacterial drug(s).
2. The hosts MIC breakpoint. A breakpoint is a chosen concentration of antibiotic which defines whether a species of bacteria is susceptible or resistant. If the MIC is less than or equal to the susceptibility breakpoint the bacteria is considered susceptible to the antibiotic. The closer the MIC is to the breakpoint the greater the chance that the antibiotic may not be clinically effective.

3. The drug's volume of distribution (Vd) in the host species.

4. Therapeutic factor (TF). MIC values are multiplied by a therapeutic factor, often between 4 and 10, in order to compensate for differences between the culture MIC and the patient's condition (eg, organism virulence, immune status of the patient, seriousness of the infection, difficulty for the drug to penetrate to the site of infection, etc).

5. Target PDC (plasma drug concentration) = MIC x TF, should be < breakpoint of the antibiotic organism combination.

6. Patient dose = PDC x Vd.

Examples of calculated dosages for *Salmonella* hepatitis in a slider are provided in Table 3.

Table 3. Hepatic biopsy reveals *Salmonella* hepatitis in a slider (*Trachemys scripta*), calculated drug doses.

Drug	Disc sensitivity	MIC (µg/ml)	Vd (L/kg)	Calculated dose[a] (mg/kg)
Carbenicillin	Sensitive	16	0.24	31 mg/kg[b]
Enrofloxacin	Sensitive	1	2.48	20 mg/kg[c]
Metronidazole	Resistant	128	0.74	758 mg/kg[d]

[a]Calculated drug doses from MIC calculated as follows: MIC (µg/ml = mg/L) x Vd (L/kg) x treatment factor (often 4-10, using 8 in this example).
[b]The calculated dose (31 mg/kg) is well below the pharmacokinetically evaluated, published dose (200 mg/kg). Therefore, use of the published dose would easily achieve high tissue levels and greater likelihood of treatment success, despite the *intermediate* disc sensitivity.
[c]The calculated dose (20 mg/kg) is greater than the pharmacokinetically evaluated, published dose (10 mg/kg). Therefore, use of the published dose would fail to achieve adequate tissue levels and result in treatment failure, despite the *sensitive* disc sensitivity. The high calculated dose may achieve adequate tissue levels at the expense of causing drug toxicity.
[d]The calculated dose (758 mg/kg) is far greater than the pharmacokinetically evaluated, published dose (20 mg/kg). Therefore, use of the published dose would fail to achieve adequate tissue levels and result in treatment failure, and is consistent with the *resistant* disc sensitivity. It would likely be impossible to achieve such high tissue levels, and attempts to do so would cause drug toxicity.

Evaluation of therapy

There are many reasons why even pharmacokinetically derived drug doses may result in inappropriate drug levels, even in the same species. The author has been surprised by how measured drug levels often deviate significantly from those of published pharmacokinetic studies, resulting in the need to adjust doses by up to 50%. Therefore, measuring drug levels in chronic cases requiring prolonged therapy is recommended.

Most, if not all, pharmacokinetic studies use healthy experimental animals. Hydration, concurrent disease, temperature, nutrition, age, gender, season and reproductive state, etc, can influence drug absorption, distribution and elimination. In addition, most pharmacokinetic investigations operate over a short time period, often a single dose, and therefore the longer term effects of accumulation or induced metabolism are seldom appreciated. Finally, compounded drugs may not act in the same way as manufactured products. For example, although not an antibiotic, compounded itraconazole consistently failed to achieve any detectable plasma levels in tortoises, whereas the commercial product produced high levels (unpublished observations). Manufactured oral ciprofloxacin contains microcapsules (hence its high cost) and therefore compounding tablets to obtain a cheaper oral suspension may result in less reliable pharmacokinetics. Compounding metronidazole to improve palatability may also risk changing absorption characteristics.

References

1. Llor C, Bjerrum L. Antimicrobial resistance: risk associated with antibiotic overuse and initiatives to reduce the problem. *Ther Adv Drug Saf.* 2014;5:229-241.

2. Weese JS, Giguère S, Guardabassi L, et al. ACVIM consensus statement on therapeutic antimicrobial use in animals and antimicrobial resistance. *J Vet Intern Med.* 2015;29:487-498.

3. Weese JS. Investigation of antimicrobial use and the impact of antimicrobial use guidelines in a small animal veterinary teaching hospital: 1995-2004. *J Am Vet Med Assoc.* 2006;228:553-558.

4. American Veterinary Medical Association. Policy statement: judicious therapeutic use of antimicrobials. Available at https://www.avma.org/KB/Policies/Pages/Judicious-Therapeutic-Use-of-Antimicrobials.aspx Accessed June 30, 2017.

5. Stone HH, Haney BB, Kolb LD, et al. Prophylactic and preventive antibiotic therapy: timing, duration and economics. *Ann Surg.* 1979;189:691-698.

6. Mathur P, Trikha V, Farooque K, et al. Implementation of a short course of prophylactic antibiotic treatment for prevention of postoperative infections in clean orthopaedic surgeries. *The Indian J Med Res.* 2013;137:111-116.

7. Avni T, Levcovich A, Ad-El DD, et al. Prophylactic antibiotics for burns patients: systematic review and meta-analysis. *BMJ.* 2010;340:c241.

8. Offenbartl K, Bengmark S. Intraabdominal infections and gut origin sepsis. *World J Surg.* 1990;14:191-195.

9. Flowers CR, Seidenfeld J, Bow EJ, et al. Antimicrobial prophylaxis and outpatient management of fever and neutropenia in adults treated for malignancy: American Society of Clinical Oncology clinical practice guideline. *J Clin Oncol.* 2013;31:794-810.

Evaluation of Gastrointestinal Activity Patterns in Healthy Rabbits Using Duplex Doppler Sonography

Trisha J Oura, DVM, Dipl ACVR,
Emi Knafo, DVM, Dipl ACZM,
Juliet Gladden, DVM, DAVECC,
Jennifer Graham, DVM, DABVP (Avian Practice), DABVP (Exotic Companion Mammal Practice), Dipl ACZM

Session #2018

Affiliation: From the Department of Clinical Sciences, Cummings School of Veterinary Medicine at Tufts University, 200 Westboro Rd, North Grafton, MA 01536, USA.

Rabbit gastrointestinal syndrome (RGIS) is one of the most common clinical disorders of pet rabbits and can be a primary or secondary condition.[1,2] Radiography has traditionally been the single most important diagnostic test for evaluation of the gastrointestinal (GI) tract of rabbits with RGIS and for monitoring response to therapy. Drawbacks of radiographic imaging in rabbits include stress to the patient, risk of injury, necessity for sedation in many cases, non-quantitative results, potential for staff radiation exposure, and inability to distinguish mechanical and paralytic ileus. Ultrasound is a common modality used in human and veterinary medicine to evaluate the gastrointestinal tract.[3,4] Duplex Doppler sonography (simultaneous B-mode and pulsed wave Doppler) of the GI tract has been used in humans, horses, and dogs to differentiate mechanical and paralytic ileus.[5-7] Our objectives were 1) to establish baseline B mode and Doppler patterns of the normal rabbit GI tract using duplex Doppler sonography and 2) to use duplex Doppler sonography to compare GI peristalsis of healthy sedate vs non-sedate rabbits.

Ten healthy non-sedate client-owned rabbits of randomly selected breed, sex, and size were used to establish baseline Duplex Doppler sonography measurements. Quantitative measurements of gastric, duodenal, cecal, jejunal, and colonic peristalsis were acquired from 3 30-second video clips of pulsed-wave Doppler (PWD) ultrasound of each region as previously described.[5-7] Simultaneous B mode ultrasonography was used to record qualitative measurements of motility in the same regions. Similar measurements were also taken on 10 healthy (randomly selected breed, sex, and size) rabbits presenting for routine ovariohysterectomy or castration before and after sedation with butorphanol (0.5 mg/kg IM; Torbugesic, Fort Dodge Animal Health), midazolam (0.5 mg/kg IM; Midazolam, West-Ward Pharmaceutical Corp), and ketamine (15 mg/kg IM; Ketamine, Putney Inc) prior to surgery.

SAS version 9.4 was used for all statistical analysis (SAS Institute, Inc, Cary, NC). Initial analysis revealed that while duodenal and jejunal segments displayed measurable peristaltic waves, the stomach, cecum, and colonic segments did not demonstrate consistent measurable activity. Standard t-tests and Wilcoxon non-parametric comparisons showed no statistically significant difference in qualitative and quantitative peristalsis between sedate and non-sedate rabbits and Pearson (parametric) and Spearman (non-parametric) correlations of the sedated measures with the non-sedated measures showed no significant correlations; however, PWD and B mode measurements were highly correlated. Results of the study suggest that duplex Doppler sonography of duodenal and jejunal segments may be suitable for non-invasive evaluation of small intestinal motility in rabbits and that the sedation protocol used in this study had no impact on measured peristaltic values.

References

1. Graham JE. Lagomorpha (pikas, rabbits, and hares). In: Miller E, Fowler ME, eds. *Fowler's Zoo and Wild Animal Medicine.* Vol 8. St. Louis, MO: Elsevier Saunders;2015:375-384.

2. DeCubellis J, Graham JE. Gastrointestinal disease in guinea pigs and rabbits. *Vet Clin North Am Exot Anim Pract.* 2013;16(2):421-436.

3. Suri S, Gupta S, Sudhakar PJ, et al. Comparative evaluation of plain films, ultrasound and CT in the diagnosis of intestinal obstruction. *Acta Radiol.* 1999;40(4):422-428.

4. Banzato T, Bellini L, Contiero B, et al. Abdominal ultrasound features and reference values in 21 healthy rabbits. *Vet Rec.* 2015;176(4):101-108.

5. Gimondo P, La Bella A, Mirk P, Torsoli A. Duplex-Doppler evaluation of intestinal peristalsis in patients with bowel obstruction. *Abdom Imaging.* 1995;20(1):33-36.

6. Mitchell C, Malone E, Sage A, Niksich K. Evaluation of gastrointestinal activity patterns in healthy horses using B mode and Doppler ultrasonography. *Can Vet J.* 2005;46:134-140.

7. An Y, Lee H, Chang D, et al. Application of pulsed Doppler ultrasound for the evaluation of small intestinal motility in dogs. *J Vet Sci.* 2001;2(1):71-74.

Conscious Computed Tomography Scanning as a Diagnostic Tool in Exotic Animal Practice

Jenna Richardson, BVM&S, MRCVS,
Tobias Schwarz, MA, Dr med vet, DipECVDI, Dipl ACVR, DVR, MRCVS,
Kevin Eatwell, BVSc (hons), DipZooMed (Reptilian), DipECZM (Herpetology), DipECZM (Small Mammals), MRCVS

Session #2064

Affiliation: From the Rabbit and Exotic Animal Department, Royal (Dick) School of Veterinary Studies. University of Edinburgh, Edinburgh, United Kingdom.

Exotic animal patients commonly present to the veterinary clinic debilitated and requiring urgent veterinary treatment. This is often due to a variety of factors including owner inexperience at recognizing signs of disease and identifying abnormal behavior, as well as the ability of the animal to mask clinical signs of disease. Particularly in prey species, handling for physical examination, when already in a compromised health state, can be detrimental and even life threatening to the patient. Understandably, performing diagnostic investigations that require manual restraint or a general anesthetic in such animals increase the risk of further deterioration or even death. While both ultrasound and radiography can be performed conscious in the majority of exotic pet species, ultrasonography requires manual restraint by the veterinary surgeon which can cause considerable stress to the patient. The quality of the radiographic images, and therefore ability to obtain a diagnosis, is dependent on the appropriate position of the patient. Conscious radiography often results in the superimposition of soft tissue and musculoskeletal structures which impedes diagnosis. The use of conscious computed tomography (CT) scanning eliminates the need for manual restraint, with image quality dependent on the stillness of the patient, as movement artefact compromises diagnostic quality. For ill animals, where lethargy and depression are common presenting signs, reduced activity levels are to be expected, increasing the chance of diagnostic quality CT images being obtained. Overall this useful, quick and repeatable modality of diagnostic imaging is extremely useful in rabbit and exotic animal veterinary practice.

References

1. Bohler A, Henninger W. Computed tomography of the rabbit head without general anaesthesia. *Wiener Tierarztliche Monatsschrift.* 2008;95(5-6):116-120.

2. De Rycke LM, Boone MN, Van Caelenberg AI, et al. Micro-computed tomography of the head and dentition in cadavers of clinically normal rabbits. *Am J Vet Res.* 2012;73(2):227-232.

3. Oliveira CR, Pijanowski GJ, Hartman S, et al. The better mousetrap (Tm): a device for computed tomography imaging of cats without general anesthesia. *Vet Radiol Ultrasound.* 2010;51(2):194.

4. Oliveira CR, Ranallo FN, Pijanowski GJ, et al. The vet mousetrap: a device for computed tomographic imaging of the thorax of awake cats. *Vet Radiol Ultrasound.* 2011;52(1):41-52.

Practice Tips for Exotic Animals

Teresa Bradley Bays, DVM, CVA, DABVP (Exotic Companion Mammal Practice)

Session #2020

Affiliation: From Belton Animal Clinic and Exotic Care Center, 1308 N Scott Avenue, Belton, MO 64012, USA and Raymore Animal Clinic, 406 W Pine Street, Raymore, MO 64083, USA.

Abstract: Every practice does things a little differently and techniques tend to evolve over time. This presentation provides examples of what to do and advice on what not to do in order to make everyday practice more successful for the betterment of our exotic animal patients.

Obesity is an increasing problem in exotic patients. Clients should be counseled early about nutrition, how much to feed, and how to provide exercise.

Collecting urine for urinalysis in small herbivores can be done by manual expression over a clean sheet of plastic when bladder is too small for ultrasound guided cystocentesis. Checking with ultrasound or radiographs first (especially for guinea pigs) to make sure there are no cystic calculi present is essential to prevent pressure forcing calculi into the trigone or urethra.

Some small mammals such as hedgehogs will urinate (and defecate) readily if placed in a small plastic critter carrier. Ferrets will often urinate and defecate on the floor in a corner of the room or if placed in the exam room sink.

Dry unshed skin (dysecdesis) on the toes and feet of reptiles can often be soaked off and should not be pulled off unless softened first. Leaving the unshed skin on may lead to necrosis of the toes and constriction of the limbs.

Feed appropriately sized food and prey items. Place animals on substrates that are not easily ingested. Do not use calcium-containing sand, which is a frequent source of intestinal obstruction in bearded dragons and leopard geckos.

Ovariectomy or ovariosalpingectomy should be considered as a preventive measure in healthy young reptiles not intended for breeding to prevent compromise caused from pre-ovulatory stasis and dystocia.

Deformities caused by malnutrition in juvenile reptiles, if not caught early, can become substantially worse as they age. Clients should be made aware of this when making decisions to treat juvenile reptiles with severe nutritional secondary hyperparathyroidism.

Dorsoventral radiographs of reptiles can be facilitated by providing a clear plastic box in front of the reptiles head to keep them from moving. Standing lateral radiographic views for reptiles allows for better visualization of air space, better definition of ileus and more information in lizards which lack a diaphragm. Otherwise superimposition of soft tissues beneath the air sacs/lung makes it difficult to assess health of the lungs and more difficult to discern the presence of ileus. It is important when taking radiographs to get the entire patient in the film in order not to miss anything. This includes getting all the way past the pelvis and entire urethra on lateral views. If taking radiographs take views of the entire patient and not just where the symptoms are present so important information for diagnosis, prognosis and treatment are not missed,

Intubating reptiles can be facilitated by having the holder push up in the dewlap area in order to elevate the trachea.

Anesthetic recovery in reptiles is facilitated by IPPV with room air using an ambubag instead of pure oxygen as they need the stimulus of CO_2 to make them breathe. Using a balloon for anesthetic bag on non-rebreathing circuit creates less resistance for small mammals and other small critters.

Even those cases with guarded to poor prognosis should be considered for full workups and surgery (if needed) as long as the client is aware of the risks, prognosis and financial commitment.

Hedgehogs can be difficult to examine unless extremely well socialized. Visualization of the ventrum, chin, neck, and feet is made easier by placing them in a clear plastic container.

Older male guinea pigs should have the penis and peri-anal folds checked for buildup of sebum, hair feces and debris. The areas should be cleaned gently with moistened cotton tipped swabs.

Enrofloxacin injections should not be used repeatedly in snakes. They are very harsh on subcutaneous tissues even if diluted.

Guinea pigs should always be on vitamin C daily, especially if ill or hospitalized. Scurvy and degenerative joint changes can occur quickly, especially if the patient is already compromised.

Be a detective: why has this sugar glider needed to be on antibiotics on and off for 2 years because of gingivitis and gingival recession? Examine and do diagnostics for the whole pet before making decisions.

Young potbellied pigs are especially prone to salt toxicity if they do not drink a lot of water or if water is deprived and then given in large quantities all at once. Wet the food and place treats like Cheerios or apple pieces in water.

Any disease process or procedure that is considered painful in other species should be considered painful in all exotic species as well. Always provide analgesics before doing any diagnostics or other procedures.

Advanced Hematology of Non-mammalians: Standardization and Consistency in Reporting Abnormal Morphologic Changes

Nicole Stacy, DVM, Dr med vet, DACVP,
Terry Norton, DVM, Dipl ACZM,
Carolyn Cray, PhD,
Kendal Harr, DVM, MS, DACVP

Session #2088

Affiliation: From Department of Large Animal Clinical Sciences, College of Veterinary Medicine, University of Florida, 2015 SW 16th Ave, Gainesville, FL 32610, USA (Stacy), The Georgia Sea Turtle Center, The Jekyll Island Authority, 214 Stable Road, Jekyll Island, GA 31527, USA (Norton), Avian and Wildlife Laboratory, Department of Pathology and Laboratory Medicine, University of Miami Miller School of Medicine, Miami, FL 33136, USA (Cray, Stacy), Urika Pathology LLC, 8712 53rd Pl W, Mukilteo, WA 98272, USA (Harr).

Morphologic changes in blood cells of any species provide valuable diagnostic information to the clinician that can be useful for diagnosis, monitoring, prognosis, and guidance for additional diagnostic testing. Although the quantitative and descriptive methods of blood cell morphologic changes in human and mammalian laboratory medicine have been well described in the literature[1] and are integrated in good laboratory practice standards, this process is lacking in non-mammalian species. Various methods have been published,[2-4] but there is great variation and thus sources for inconsistencies in reporting among evaluators. Morphologic changes in blood cells of non-mammalian species include the following: erythroid regeneration, hypochromasia, poikilocytosis, inclusions, and hemoparasites in red blood cells; heterophil/neutrophil toxicity, heterophil/neutrophil left-shifting, reactive changes in lymphocytes and monocytes, inclusions, presence of melanomacrophages, infectious agents, and neoplasia in white blood cells; and activation and phagocytosis in thrombocytes. As an example, there is great variability in reporting heterophil toxicity,[2-4] which is a critical morphologic change in sick patients and helpful for monitoring during treatment. The possibilities of reporting toxicity range from a subjective evaluation (eg, mild, moderate, marked) to reporting percentages, absolute numbers, or a scale of 1+ to 3+ or 1+ to 4+ with information on how many cells are affected. In contrast to mammalian neutrophil/heterophil toxicity that occurs orderly depending upon the severity, there is great variability in non-mammalians associated with toxic change. Standardization of the description of blood cell morphologic changes in non-mammalians will be useful for comparison and consistency of laboratory results.

References

1. Harvey JW. *Veterinary Hematology: A Diagnostic Guide and Color Atlas.* St. Louis, MO: Elsevier Saunders;2012.

2. Thrall MA, Weiser G, Allison R, Campbell T. *Veterinary Hematology and Clinical Chemistry.* 2nd ed. Ames, IA: John Wiley & Sons Inc;2012.

3. Campbell TW. *Exotic Animal Hematology and Cytology.* 4th ed. Ames, IA: Wiley-Blackwell;2015.

4. Strik, NI, Alleman AR, Harr KE. Circulating inflammatory cells. In: Jacobson ER, ed. *Infectious Diseases and Pathology of Reptiles.* Boca Raton, FL: CRC Press;2007:167-218.

Neoplasia in Reptiles: A Retrospective Study (2010-2015)

Tom Hellebuyck, Dr med vet, DipECZM (Herpetology),
Jules Simard, med vet,
Sylvie Pirson, med vet,
Leslie Bosseler, med vet, Dipl ECVP,
An Martel, Prof Dr med vet, DipECZM (Wildlife Population Health)

Session #2074

Affiliation: From the Department of Pathology, Bacteriology and Avian Diseases, Faculty of Veterinary Medicine, Ghent University, B-9820 Merelbeke, Belgium.

This study documents the occurrence of neoplastic disorders in reptiles presented at our department from January 2010 to December 2015 based on antemortem diagnostic investigations. Neoplasia was most common in lizards (48.6%), followed by snakes (34.3%) and chelonians (17.1%). In none of the cases multiple types of neoplasia were diagnosed in the same individual. A multicentric appearance was seen for squamous cell carcinomas (SCC), chromatophoromas and papillomas and the presence of metastatic tumors was diagnosed in 25.7% of the cases. The latter cases mostly presented with end stage or recurrent neoplastic disease. Primary neoplasms of the integument were observed most commonly (51.4%). Gastrointestinal and liver neoplasia was seen in 28.6% of the cases and neoplasia of the musculoskeletal system and the reproductive tract was observed in 5.7% and 5.6% of the cases, respectively. Based on the results of the present study, bearded dragons seem to be predisposed to the development of cutaneous SCC and gastric neuroendocrine carcinoma and basal cell carcinomas seem to occur most commonly in geriatric *Testudo hermanni*. Remarkably, 2 hepatocellular carcinomas were demonstrated in 2 Chinese rat snakes (*Orthriophis taenurius*). In general, complete tumor resection proved to be the treatment of choice. Previously published retrospective studies about reptile neoplasia mostly originate from data collected by specialty diagnostic services.[1,2] The present study offers an overview of the most common neoplastic disorders that may be encountered by the clinical practitioner during day-to-day practice based on an antemortem diagnostic approach.

References

1. Garner M, Hernandez-Divers S, Raymond J. Reptile neoplasia: a retrospective study of case submissions to a specialty diagnostic service. *Vet Clin North Am Exot Anim Pract.* 2004;7(3):653-671

2. Sykes JM, Trupkiewicz JG. Reptile neoplasia at the Philadelphia Zoological Garden, 1901-2002. *J Zoo Wildl Med.* 2006;37(1):11-19.

Gap in pagination due to unavailable material.

Pages 52-79

Section 5

ARAV Nutrition

Nelson Bricker, DVM and
David Hannon, DVM, DABVP (Avian Practice)
Session Chairs

Sponsored By:

UMIAMI
AVIAN & WILDLIFE LABORATORY

Trace Element Serum Levels in *Testudo* spp.

P. Koelle, DVM, Dr med vet, PrivDoz, Cert Spec for Reptiles, Cert Spec for Fish (including Ornamental Fish), Cert Spec for Nutrition Consultation (Pets),
H. Stephan, DVM,
S. Öfner, DVM,
T. Hollandt, DVM,
S. Caron,
J. M. Ballouard,
S. Reese, DVM, Dr med vet, PrivDoz, Cert Spec for Anatomy, Histology and Embryology, Cert Spec for Information Technology

Session #2057

Affiliation: From Clinic for Small Animal Internal Medicine of Ludwig-Maximilians-University Munich, Veterinaerstrasse 13, D-80539 Muenchen, Germany (Koelle, Stephan), Reptile Rescue Center Munich, Kaulbachstr. 37, D-80539 Muenchen, Germany (Öfner, Hollandt), Station d'Observation et de Protection des Tortues et de leurs Milieux Chelonian Conservation and Research Center (CRCC), BP24, F-83590 Gonfaron, France (Caron, Ballouard), Lehrstuhl fuer Anatomie, Histologie und Embryologie, Veterinaerstrasse 13, D-80539 Muenchen, Germany (Reese).

Trace elements are essential for physiologic function of the organism in all animals, including reptiles. Nutritional requirements of trace elements are not known in reptiles. Trace elements of chelonians were examined in serum samples of (probably not healthy) pet tortoises,[1] for evaluation of environmental pollution in sea turtles,[2] and from material of carcasses.[3] The aim of the study was to evaluate reference serum parameters of iron, zinc, iodine, manganese, copper, and selenium in *Testudo* spp. Blood samples were obtained from 115 tortoises (63 male and 52 female specimens) from the dorsal coccygeal vein. Blood was collected from 69 pet tortoises (*Testudo* spp.) in Germany and from 46 free-ranging *Testudo hermanni* in France. Zinc, manganese, selenium, and copper values were comparable to values cited in the literature and also to values reported from herbivorous mammal species, such as horses.[1,4] No statistically significant differences were detected between samples from tortoises originating from Germany or France, with the exception of copper values, which were higher in the French tortoises. Zinc values were higher in male tortoises than in females.

Acknowledgments: This study was sponsored by JBL (Neuhofen, Germany).

References

1. Mueller A, Raich S, Bauder S, Altherr B. Trace elements in reptiles in comparison to mammals. *Proc 1st Int Conf Avian Herpetol Avian Med.* 2013;174-175.

2. Maffucci F, Caurant F, Bustamante P, Bentivegna, F. Trace element (Cd, Cu, Hg, Se, Zn) accumulation and tissue distribution in loggerhead turtles (*Caretta caretta*) from the Western Mediterranean Sea (southern Italy). *Chemosphere.* 2005;58(5):535-542.

3. Kienzle E, Kopsch G, Koelle P, Clauss, M. Chemical composition of turtles and tortoises. *J Nutr.* 2006;136(7):2053S-2054S.

4. Laboklin. Reference values. Available at: http://oldsite.laboklin.de/pages/html/de/leistungsspektrum/referenzwerte/referenzwert_hdktz.htm Accessed June 30, 2017.

You Are What You Eat! Evaluating a Commercial Bearded Dragon Diet

Laure Molitor, MS, DVM,
Ian Konsker, DVM,
Mark Mitchell, DVM, MS, PhD, DipECZM (Herpetology)

Session #2092

Affiliation: From the Department of Veterinary Clinical Sciences, School of Veterinary Medicine, Louisiana State University, Skip Bertman Drive, Baton Rouge, LA 70803, USA.

Bearded dragons (*Pogona vitticeps*) are one of the most popular pet reptiles kept in captivity. The purpose of this study was to generate evidence-based nutritional data on the health of bearded dragons, using an experimental semi-moist diet with the goal of creating a nutritionally balanced diet and decreasing the prevalence of secondary nutritional hyperparathyroidism. Bearded dragons were randomly divided into 2 groups: treatment (12 dragons) and control (12 dragons). Dragons assigned to the treatment group were provided a semi-moist commercial diet, while dragons in the control group were provided non-supplemented crickets. The total weight of the semi-moist gel diet or crickets offered was 2% of the dragon's body weight daily. Over the course of the study, body weights were measured weekly. MicroCT scans were collected at baseline (day 0), 45 days, 90 days, and 120 days after initiating the study. Houndsfeld unit density collected from the mandibular cortical bone was used to compare the 2 groups. The semi-moist diet had 3.6% calcium, while the cricket diet had 0.11% calcium. The calcium to phosphorus ratios for the diets were 10:1 and 1:10, respectively. Vitamin A concentration on an as is basis in the semi-moist and cricket diets were 6000 IU/kg and < 1000 IU/kg, respectively.[1] Finally, vitamin D concentrations on an as is basis in the semi-moist and cricket diets were 3400 IU/kg and < 250 IU/kg respectively.[1] While the dragons consumed both diets, the semi-moist diet had calcium, vitamin A, and vitamin D concentrations that were considered more physiologically appropriate for dragons.

References

1. Finke MD. Gut loading to enhance the nutrient content of insects as food for reptiles: a mathematical approach. *Zoo Biol.* 2003;22:147-162.

Testing the Delivery of Essential Vitamins and Calcium to Geckos via Crickets

Laure Molitor, MS, DVM,
Ian Konsker, DVM,
Mark Mitchell, DVM, MS, PhD, DipECZM (Herpetology)

Session #2093

Affiliation: From the Department of Veterinary Clinical Sciences, School of Veterinary Medicine, Louisiana State University, Skip Bertman Drive, Baton Rouge, LA 70803, USA.

Leopard geckos are one of the most popular pet reptiles kept in captivity. The purpose of this study was to generate evidence-based nutritional data using an experimental cricket gutloading diet with the goal decreasing the prevalence of secondary nutritional hyperparathyroidism by creating a nutritionally balanced diet for these animals. Leopard geckos were randomly divided into 2 groups: treatment (12 geckos) and control (12 geckos). Geckos assigned to the treatment group were provided crickets supplemented with the experimental diet, while geckos in the control group were provided non-supplemented crickets. The total weight of crickets offered daily was 2% of the leopard gecko's body weight. Over the course of the study, body weights were measured weekly. MicroCT scans were collected at baseline (day 0), 45 days, 90 days, and 120 days after initiating the study. Houndsfeld unit density collected from the mandibular cortical bone was used to compare the 2 groups. The experimental cricket diet had 5.9% calcium dry matter, while the non-supplemented crickets had 0.11% calcium. The calcium to phosphorus ratios for the diets were 10:1 and 1:10, respectively. Vitamin A concentration on an as is basis in the experimental cricket diet and non-supplemented crickets was 5000 IU/kg and < 1000 IU/kg.[1] Vitamin D concentrations on an as is basis in the experimental diet and fasted crickets were 1000 IU/kg and < 250 IU/kg respectively.[1] While the geckos consumed both diets, the experimental cricket diet had calcium, vitamin A, and vitamin D concentrations that were considered more physiologically appropriate for geckos.

References

1. Finke MD. Gut loading to enhance the nutrient content of insects as food for reptiles: a mathematical approach. *Zoo Biol.* 2003;22:147-162

Section 6

ARAV Reproduction Bonus Session

Nelson Bricker, DVM and
David Hannon, DVM, DABVP (Avian Practice)
Session Chairs

Salpingotomy in a Shingleback Skink with Subsequent Successful Parturition

Alicia McLaughlin, DVM, and
Anneliese Strunk, DVM, DABVP (Avian Practice)

Session #2096

Affiliation: From Center for Bird and Exotic Animal Medicine, 11401 NE 195th St, Bothell, WA 98011, USA.

A 5-year-old female shingleback skink (*Tiliqua rugosa rugosa)* presented for anorexia, lethargy, and coelomic distention 3 weeks after passing a healthy offspring. Radiographic imaging confirmed the presence of 3 fetuses with calcified skeletons within the coelom. A salpingotomy was performed to remove the fetuses, which were macerated, necrotic, and surrounded by caseous debris. The oviduct was inflamed and friable. Following surgical debridement and flushing with sterile saline, the oviductal incision was sutured closed. The patient received analgesics and an extended course of antibiotics following surgery, and proceeded to make an uneventful recovery. One year later, the patient produced 3 healthy offspring without medical or surgical assistance, indicating a return to normal reproductive function. The cause for fetal death in this case remains unknown. It is possible that bacterial salpingitis led to fetal death, or could have developed secondary to fetal death and subsequent necrosis. Salpingotomy should be considered as an alternative to ovariectomy/salpingohysterectomy in cases of oviductal disease in valuable breeding animals experiencing dystocia.

References and Suggested Reading

1. Di Girolamo N, Mans C. Reptile soft tissue surgery. *Vet Clin North Am Exot Anim Pract.* 2016;19(1):97-131.

2. Divers SJ. Reptile coeliotomy: a vital technique to master. *Proc Annu Conf North Am Vet Comm.* 2006; 1614-1618.

3. Chiodini RJ, Sundberg JP. Cesarean delivery in the snake. *Small Anim Clin.* 1980;75(10):1605-1608.

4. Munns SL, Daniels CB. Breathing with big babies: ventilation and oxygen consumption during pregnancy in the lizard *Tiliqua rugosa. Physiol Biochem Zool.* 2007;80(1):35-45.

5. Fergusson B, Bradshaw SD. Plasma arginine vasotocin, progesterone, and luteal development during pregnancy in the viviparous lizard *Tiliqua rugosa. Gen Comp Endocrinol.* May 1991:82(1):140-151

6. Fergusson B, Bradshaw SD. In vitro uterine contractions in the viviparous lizard *Tiliqua rugosa*: effects of gestation and steroid pre-treatment. *Gen Comp Endocrinol.* 1992;86(2):203-210.

ARAV
ASSOCIATION OF REPTILIAN AND AMPHIBIAN VETERINARIANS

Section 7

ARAV Diagnostics

Heather Bjornebo, DVM and
Teresa Bousquet, DVM
Session Chairs

Sponsored By:

UMIAMI
AVIAN & WILDLIFE LABORATORY

Anatomy of the Gastrointestinal Tract and Transit Time of Gastrografin® in Inland Bearded Dragons (*Pogona vitticeps*)

Karina Mathes, DVM, Dr med vet, Cert Spec Reptiles, DipECZM (Herpetology),
Katharina Radelof, DVM,
Elisabeth Engelke, DVM, Dr med vet,
Christiane Pfarrer, DVM, Dr med vet, Prof,
Michael Fehr, DVM, Dr med vet, DipECZM (Small Mammals), Prof

Session #2054

Affiliation: From Clinic for Small Mammals, Reptiles and Birds, University of Veterinary Medicine Hannover, Foundation, Hannover, Buenteweg 9, Germany (Mathes, Radelof, Fehr), Institute of Anatomy, University of Veterinary Medicine Hannover, Foundation, Hannover, Germany (Engelke, Pfarrer).

The aim of the study was to determine the radiographic gastrointestinal anatomy of adult healthy inland bearded dragons (*Pogona vitticeps*) and the contrast transit times of orally administered meglumine-amidotrizoate in this species. In this study, 21 healthy adult bearded dragons (12 females and 9 males) were orally administered Gastrografin® (meglumine amidotrizoate) at a dose of 5 ml/kg. Dorsoventral and laterolateral radiographs were performed after 0, 15, 30, 45 minutes as well as 1, 2, 3, 4, 5, 6, 8, 10, 12, 24, 30, and 36 hours after oral contrast medium administration. Gastrografin® was administered orally via syringe directly into the oral cavity and voluntarily swallowed by all subjects. Gastrografin® was sufficient to illustrate the gastrointestinal tract in each of the 21 bearded dragons. In general, the evaluation of every part of the digestive tract was feasible with dorsoventral and laterolateral radiographs taken 30 minutes, 6 hours, and 30 hours after oral administration of the contrast medium. The combination of consecutive dorsoventral radiographs and laterolateral horizontal-beam radiographs showed the best results in this study. Oral administration of Gastrografin® (meglumine amidotrizoate) at a dose of 5 ml/kg appears to be a safe technique in all 21 tested lizards. The use of this contrast medium at the given dose seems to be a reliable technique for the radiographic evaluation of the gastrointestinal tract in inland bearded dragons within 36 hours.

Why Stick a Needle in the Heart When a Jugular Will Do? Developing a Consistent Approach for Jugular Venipuncture in the Snake

Sean M. Perry, DVM,
Stephen Joslyn, BSc, BVMS, DipECVDI, MRCVS,
Mark A. Mitchell, DVM, MS, PhD, DipECZM (Herpetology)

Session #2080

Affiliation: From Louisiana State University, Veterinary Clinical Sciences, Skip Bertman Drive, Baton Rouge, LA 70810, USA (Perry, Mitchell), Vet CT, 185 High Street, Fremantle, Western Australia, 6160, Australia (Joslyn).

Venipuncture in snakes has traditionally been limited to 2 sites: the ventral coccygeal vein and heart. Ventral coccygeal venipuncture can be difficult in some snakes due to their short tails (eg, ball pythons), and while commonly done, cardiocentesis is not without its potential complications (eg, lacerated heart or great vessel). Jugular venipuncture is commonly used in chelonians and lizards, and has been reported in snake species; however, a consistent approach to performing venipuncture at this vessel in snakes has not been objectively described in the literature. An observational study was performed to determine external landmarks in ball pythons that correspond with internal vascular anatomy using computed tomography (CT). Six ball pythons (2 males, 4 females) were enrolled in the study. Animals were sedated with alfalaxone 15 mg/kg intramuscularly to collect the external measurements and for CT imaging. A GE 16 slice LightSpeed CT-scanner unit was used to characterize the vascular anatomy of the animals. To evaluate the internal vascular anatomy, 2 ml/kg of iohexol was administered. External morphometric data collected during the study was as follows: heart snout length, mean: 30.36 ± 9.0 cm, min- max, 19.10- 40.5 cm; heart length (cm), mean: 3.6 ± 0.9 cm, min-max, 2.4-4.5 cm; and heart length (scales), 6.9 ± 1.9 scales, min-max [5-10.5 scales]. The iohexol contrast was useful in identifying the jugular vessels in ball pythons.

Determining Baseline Corticosterone Concentrations in Captive and Wild-Caught Veiled Chameleons (*Chamaeleo calyptratus*): How Stressed Do These Animals Really Get?

Ian Konsker, DVM,
Sean Perry, DVM,
Rory Fuller, BS,
Michael Romero, BA, MS, PhD,
Mark Mitchell, DVM, MS, PhD, DipECZM (Herpetology)

Session #2081

Affiliation: From the Department of Veterinary Clinical Sciences, School of Veterinary Medicine at Louisiana State University, Skip Bertman Drive, Baton Rouge, LA 70803, USA, Department of Biology, School of Arts and Sciences at Tufts University, 165 Packard Avenue, Medford, MA 02155, USA.

Evaluating the effects of stress on animals in captivity has become an important topic due to animal welfare concerns for captive reptiles. To date, evidence-based studies measuring the effects of stress on captive reptiles are limited. The purpose of this study was to determine if corticosterone concentrations could be used to measure stress in captive and wild veiled chameleons (*Chamaeleo calyptratus*). To accomplish this, it is first necessary to determine baseline corticosterone concentrations in these two populations of lizards. Because handling can affect corticosterone concentrations, the first objective of the study was to determine the time interval in which handling alters corticosterone concentrations. Captive veiled chameleons were used to assess the effect of handling and restraint on corticosterone concentrations. To test this, 12 captive veiled chameleons were used. Blood samples were collected from either the ventral coccygeal vein or jugular vein immediately after getting hands on the animal (t_0; 0-3 minutes), 3 minutes (t_3; 3-6 minutes), 6 minutes (t_6; 6-9 minutes), 9 minutes (t_9; 9-12 minutes), and 30 minutes (t_{30}; 27-30 minutes) after contact. Twelve wild veiled chameleons were collected from invasive populations in Florida and sampled identically. Corticosterone concentrations were compared between both wild and captive populations. A Shapiro-Wilk test was used to evaluate the distribution of the data. A general linear model for repeated measures was used to determine the rise in corticosterone associated with handling. The Bonferroni test was used to characterize differences between sampling times. Similar to other species, corticosterone concentrations rise in veiled chameleons following handling.

Effects of Furosemide and Fluid Therapy in Bearded Dragons (*Pogona vitticeps*)

Lily Parkinson, DVM,
Christoph Mans, Dr med vet, Dipl ACZM

Session #2129

Affiliation: From the Department of Surgical Sciences, School of Veterinary Medicine, University of Wisconsin, 2015 Linden Drive, Madison, WI 53706, USA.

Assessing dehydration and adequately rehydrating a reptile patient remains difficult and understudied. In addition, the use of furosemide, a loop diuretic, in reptiles, which do not possess loops of Henle, is controversial. In the first phase of this study, the effects of furosemide (5 mg/kg and 10 mg/kg administered every 12 hours for up to 4 doses) were evaluated in a randomized, controlled, complete cross-over design. Body weight, uric acid, packed cell volume, total solids, total protein, and osmolarity were measured at baseline and once the bearded dragon was dehydrated. In the second phase of this study, bearded dragons were experimentally dehydrated using furosemide and then administered different electrolyte solutions (PlasmaLyte A, lactated Ringer's solution, and Reptile Ringer's 1:1 D5W and PlasmaLyte A) in a randomized complete-cross-over design. Fluids were administered as a single subcutaneous bolus as 5% of the dragon's original body weight. Changes in the previously measured parameters were measured 4 and 24 hours post-fluid administration.

Gap in pagination due to unavailable material.

Pages 92-108

Section 10

Honeybee Medicine

Nicole Johnson, DVM and
Dan Johnson, DVM, DABVP (Exotic Companion Mammal Practice)
Session Chairs

Tools of the Trade: The Honeybee House Call

Susan Horton, DVM

Session #2155

Affiliation: From Chicago Exotics Animal Hospital, 3757 W Dempster St, Skokie, IL 60056 USA.

Abstract: Bee hive components will be discussed. Tools for your house call kit will be covered. In addition, honeybee anatomy will be presented.

Introduction

The tools that you will need to work with and be familiar with will be addressed. Though it is important to know what each tool is, keep in mind that most bee keepers (apiarists) will prefer that you use their equipment. This prevents contamination from other bee yards (apiaries). Knowing what the components of the hive are will help you with your site examination and understanding what the bee keeper is saying.

Hive Components

The bee hive is made up of several layers of components. Each component has a specific function.

Hive cover

Outer cover: This telescoping cover sits with its edges over the sides of the top super to protect the hive.

Inner cover: This cover creates a dead air space for insulation from heat and cold.

Small or medium supers

Bees store their extra honey in these for the bee keeper to harvest.

Medium or large supers or hive bodies

Brood chambers are the bees' living quarter. The queen lays eggs in these chambers and brood is raised. Honey also is stored for bees' food.

Queen excluder

The excluder keeps the queen bee in the brood chambers as she is too large to pass through the it.

Bottom board

This forms the hive floor. It may be solid or mesh. It helps to keep pests out.

Entrance reducer

The entrance reducer keeps pests out.

Hive stand

The stand supports the hive off the ground to keep the hive dry and to insulate the hive.

Tool Box

Nitrile gloves, puncture proof

Hood and jacket

Shoe covers; sterilize your boots and shoes after each visit.

Hive tool, bee owner may have one for you to use and probably prefers

Benadryl

EpiPen (for me)

Bee Keeper Possessions

Smoker

Frame grabber

Their own tools

Testing Equipment

Culturettes

Mason jar with mesh lid, white tub, powdered sugar, 100-ml measuring cup

Or a mite testing kit

Tray to work on

Cotton tip applicators

Microscope, cover slips, and slides

Nonfat dried milk, sterile water, and some test tubes or

Vita AFB and EFB test kits

Honeybee Anatomy

Bees are quite different from what we all normally work with. Knowing a few general aspects of their anatomy can help you figure out what is physically wrong.

Head

Compound eyes and ocelli

Antennae (topochemical olfactory sense, CO_2 receptors, moisture levels, taste receptors, Johnston's organ, flight speed)

Mouth parts (mandible and proboscis)

Thorax

Spiracles for letting in air

2 pair of wings-hooked together

3 pair of legs

Abdomen

Tracheal system

Stinger-venom gland; alarm pheromone released to mark victim

Digestive tract

- *Malipighian tubules:* function like kidneys
- *Esophagus:* crop/honey sac, large intestine, small intestine, rectum

Wax glands

Exoskeletion

Sensory hairs, hairs that hold pollen, pollen basket for pollen collection

Mandibular glands

Queens produce pheromone

Nurse bees produce either royal jelly or worker jelly

Hypopharyngeal glands

Protein rich secretions in young nurses (brood food)

Invertase (enzyme to break down sucrose into fructose and glucose in foragers

Reproduction

Virgin queen, spring time, occurs in flight 7-17 times

Queen stores enough sperm for 2-5 years

References

1. Vidal-Naquet N. *Honeybee Veterinary Medicine: Apis mellifera* L. Sheffield, UK: 5m Publishing;2015:9-28.

2. Flottum K. *Backyard Beekeeper: An Absolute Beginner's Guide to Keeping Bees.* 3rd ed. Beverly, MA: Quarry Books;2014:16-48.

Beekeeping Basic Biology and Diseases of the Hive

Kay A. Backues, DVM, Dipl ACZM

Session #2161

Affiliation: From the Tulsa Zoo, 6421 E 36th St N, Tulsa, OK 74115 USA.

Abstract: The European honeybee, *Apis mellifera* is a eusocial insect. The individual bee is not the organism; it serves the hive or super organism and cannot survive outside its relationship with the hive. In describing bees we use the term "animal" to refer to both the individual bee and the hive. Like all organisms, honeybees and the hive have their infectious, parasitic, and environmental diseases. Veterinarians interested in bee keeping and or working with beekeepers to provide care need to start with a basic understanding of the hive and its annual cycle as well as disease states. Knowing when and if human intervention is necessary can differentiate the successful beekeeper from the dabbler who owns bees for a few years and then gives up. An overview of the annual cycle of the beehive as well as introduction to significant diseases follows.

Presented at NAVC Conference, 2017.

Annual Cycle of the Hive (A Very Simplistic Overview)

The following statements are generalizations. They describe beehive activities for a temperate climate with 4 seasons.

January

During cold weather the bees cluster to keep warm in the center of the hive; they eat stored honey: cold bees slowly move inwards to avoid freezing, and center bees move outward to take their place. Bees at the center of the cluster twitch to generate heat at the center of the cluster. The queen at the center of clustered bees begins to lay more eggs as daylight lengthens in preparation for the coming spring. On warm sunny days above 50 °F foraging bees will leave the hive to look for anything to eat as well as to defecate.

February and March

The cluster continues to feed off of its honey reserves. On warm days > 50 °F workers go on foraging trips for early blooming flowers. If severe cold weather strikes or the honey stores deplete, this is a critical season for hive starvation and death. The queen's egg laying increases dramatically to produce more workers for the upcoming spring nectar flow. When the brood is being reared foragers should be returning with pollen on their legs. Drone, male, bee eggs also are laid by the queen starting this time of year. Their job is to fly outside the hive and congregate in mating groups in hope of finding a virgin queen on its nuptial flights throughout the swarming or reproduce season to come.

April and May

Egg laying by the queen peaks as the hive approaches its zenith population in late spring early summer of up to 40-60,000 bees. Spring flowers offer nectar and pollen sources for the energy and protein, respectively, needed to store honey and raise baby bees. This is the season to swarm. Swarming describes the reproductive function of the hive. A group of bees leaves the hive with the old or new virgin queen to start a new hive somewhere else;

it can be triggered by crowding but is a normal instinct even in hives with sufficient space. The term is not used to describe various flying insects in any other manner. The beekeeper should monitor the spring/summer hive for sufficient room to avoid crowding and prevent swarming as much as possible. Why prevent swarming? The hive is severely weakened by swarming and may lose 50% or more of its bees. Swarming is normal but one of the best options to prevent crowding and losing bees to swarms is to split the hive (artificial swarm) and set this split up to become a new full sized colony. One can and should add special 'super' boxes to the top of the main hive where bees will store the honey that the beekeeper will take to prevent swarming associated with crowding.

Swarming bees means that there are virgin queens being produced; they leave the hive to find breeding areas where drones congregate to mate. These areas are in the air and are found by pheromone detection by both queen and drones. After its nuptial flight(s) a virgin queen returns to its hive as a mated queen and starts to lay eggs; it will never mate again. The bee may live for up to 7 years but 3-4 is more typical, its most fecund years are the first 1-3 years of life. Most beekeepers keep young queens of only 1-2 years of age to maximize egg and thereby bee and honey production.

June and July

With the heat of summer and the main nectar flow is winding down, egg laying by the queen will begin to slow as nectar becomes more scarce. Swarming also tapers off as new hives will not have time to store enough food for winter. Young bees are still raised but the maximum eggs laid /day should be slowing down from a peak of 2000/day. In some areas honey can be harvested in July from 'super' boxes placed on top of the main brood chamber in the spring. Not all the honey is taken and by harvesting only what is in these top boxes the beekeeper ensures that there is ample honey for the bees survive winter.

August and September

Honey is harvested by the beekeeper. Nectar flow from flowers slows and bees harvest water to keep the hive cool. Workers use evaporative cooling by fanning their wings over the water they have spread at the entrance up and out the top of the hive. The hive population reaches a summer nadir. As cooler weather arrives a fall flower bloom and nectar flow occurs that is vital for winter survival. A smaller boost in bee numbers will accompany this bloom. A full, large hive box of capped and stored honey should be present above the brood box by the end of fall or the beekeeper should supplement the hive's stored food with 2:1 sugar water.

October and November

The foraging bees leave the hive on sunny days when the temperatures exceed 50F and the sun is shining. A few late fall blooming plants provide some pollen and nectar. The bees take 'cleansing flights' to defecate whenever weather permits through the winter. The beekeeper feeds a hive that is light on food stores but only visits the hive as necessary, as it is stressful to disrupt a cluster of bees trying to stay warm.

December

The hive clusters at any time temperatures fall below 50-55F to keep the center bees and queen warm. Egg laying may stop or be at almost undetectable levels. Winter survival is based on having a large healthy population of worker bees going into the fall to keep the cluster large and its center warm and sufficient stored honey to prevent starvation.

Honeybee Diseases

This is a simplified overview of common diseases known to affect the European honeybee or its hive. It is not comprehensive. As with any living organism diseases can result from bacterial, viral, parasitic, and fungal causes. Disease can also be spread by poor husbandry practices and hygiene. Often treatment is not a medication but a manipulation of the hive and its environment.

Bacterial diseases

American foulbrood (AFB): This is the most important bacterial disease that affects bee larvae. AFB is caused by spore-forming *Paenibacillus larvae* ssp. *larvae* (formerly *Bacillus larvae).* The detection of AFB in the hive can be a death sentence for the hive. The spores can be brought to the hive by bees that drift in from a different hive or bees form the hive that have robbed a sick hive. The spores contaminate food that is fed to the larval bees. Larvae from 1-3 days old are the most sensitive to infection. Spores hatch in the gut of the larva and the vegetative form of the bacteria grows and kills the larva. Bees remove dead brood which is contaminated with the bacterial spores and thus spread the spores all over the hive. The spores cannot be removed from the hive and are essentially immortal, remaining viable for decades. Affected dead larvae are coffee brown, liquefied and give off a distinctive rotting smell. A simple field test to detect AFB is to touch a small stick into to the rotted mass in the cell; in cases of AFB the material is ropey and can be stretched out of the cell. Hive side tests can be purchased to confirm the diagnosis and differentiate it from other brood diseases such as European foulbrood. Also, samples of dead larva can be sent to university testing labs that specialize in bee diseases. The classic sign in the hive is depressed capped brood cells with perforations in them. If the larva lived long enough to pupate, the pupa's tongue is often found stuck to the roof of the brood cell.

Treatment will only temporarily mask signs of the disease as the spores persist in the hive indefinitely. Antibiotics labeled for use in the beehive directed at European foulbrood and AFB are available but their use is controversial. Prophylactic treatment to prevent AFB and EFB are no longer recommended due to the risk of developing resistance. Terramycin resistant AFB is known to occur. More importantly, if AFB is detected it is likely to spread to other hives. All hives in the apiary should be tested, but the suspected infected hive tested last to prevent contaminating other hives. The confirmed infected hive should be sealed and burned so that no bees can spread the disease and none of the infected hive parts can be reused. In many states AFB is a reportable disease to the state agricultural division and the state may require the burning affected hives. A few state agricultural offices offer irradiation of contaminated hive equipment, hives, frames, and woodenware. The bees should be destroyed to prevent the spread of the disease.

European foulbrood (EFB): Larvae become infected with European foulbrood when they consume brood food that contains the bacteria, *Melissococcus plutonius*. The bacteria reproduce in the gut and compete for the food that is fed by nurse bees to the larva. Depending on the level of infection and amount of available food, the infected larva will either survive or die. The larvae that survive go on to defecate and pupate, which leaves bacteria on the combs that can be infective for years, even though this bacteria does not produce spores. Surviving larvae will become adults with generally lower weight and delayed pupation when compared to their uninfected sisters. A strong hive with adequate numbers of nurse bees can reduce larval death and observed symptoms. With adequate amounts of food, larvae are more likely to survive. This explains why expression of the disease can change sporadically year to year, and season to season. EFB is more likely to be seen during times of the year when food is less available and bee numbers are low. A strong hive with a strong queen who is laying well is the best defense. Treatment with antibiotics will help stop the disease but a strong hive and queen are better preventatives.

Differentiating between AFB and EFB: Smell is one way to differentiate between AFB and EFB. AFB smells rotten and sulfurous. EFB smells more sour or not detectable. Appearance of the affected larvae is another way to differentiate. AFB larvae are coffee-colored and dark, ropey, dry-to-brittle scale in the cell. EFB larvae are dull yellow-white to brown, dry-to-rubbery scale in the cell, not as ropey with the stick test.

Hive side diagnostic tests are available. Diagnostic kits are Lateral Flow Devices that detect specific antigens that bind to antibodies in the device. Separate kits are manufactured for EFB and AFB, since two different antigens are involved. Be sure to use the correct kit for the disease suspected. Order from a bee supply catalog. Send samples of affected larvae to a university lab or the US Department of Agriculture.

Treatment of EFB: Treatment may soon require a prescription from a licensed veterinarian. Prophylactic annual treatment for EFB is not recommended! Oxytetracycline HCl soluble powder (Terramycin) is the only product labeled for the control of EFB. It is important to follow label instructions and dose correctly. Terramycin is also registered for use in controlling AFB, although AFB resistance to this drug has been observed in the United States. Infected hives with AFB should be considered for destruction due to the long survival of spores in the hive. There is no evidence of EFB becoming resistant to Terramycin in the United States to date. Use of Terramycin as a precautionary, or prophylactic, method to prevent EFB in non-symptomatic colonies, even in infected apiaries, is not recommended. Keeping strong hives, with young and fertile queens is better than treatment.

Annual frame replacement as a preventive measure for brood diseases: It is recommended to replace 5-15% of the brood comb each year to prevent the excessive buildup of chemicals, bacteria, and spores. This is an excellent integrated pest management (IPM) strategy to avoid excessive chemical use. Older frames should be discarded so that bees cannot access them. Fresh new foundation is provided for the bees to draw new wax comb on in the early spring. This is also a good measure to help prevent swarming as young bees have a place to make wax comb in their hive.

Parasitic diseases

Varroa mite: Varroa destructor appeared in European honeybees in the United States in the 1990s and led to a significant die-off of hives. The Asian honeybee, *Apis cerena*, is the natural host the Varroa mite and is resistant to significant disease from it, but Varroa is considered one of the most significant diseases of European honeybee colonies. Without treatment hives can weaken and die within 1-3 years. Heavy infestations lead to a weakened hive late in the season that likely will not survive the winter. Spread of deformed wing virus is a telltale sign of heavy infestation of Varroa. Transmission of Varroa infestation is likely from infected bees drifting to other hives, or weak, infected hives being robbed by other bees: they steal the honey but also bring the mite back to their own hives.

The life cycle is depicted by the mated female Varroa mite riding on the adult bee and drinking its hemolymph (the equivalent of its blood). After feeding it enters an open brood cell that contains a larva and lays its eggs, which quickly hatch and feed on the larvae as it pupates into an adult bee. When the bee emerges from its cell the new adult Varroa mite females emerge to attack more bees and larvae. Varroa mites seek out and preferentially enter drone larval cells, which take 2 extra days to pupate and thus give the mites more time to develop compared to worker cells. This habit can be exploited with IPM strategies.

Treating the mites kills them but leaves toxic residues in the beeswax; the mites also quickly become resistant to treatment. Instead, beekeepers are learning to live with the mites and try to help the beehive manage the mites to tolerable levels. Chemicals still can be used but sparingly. Newer more bee-friendly treatments have been introduced that lower mite populations. The use of sacrificial drone comb takes advantage of the mites' preference for drone brood. Likewise screened bottom boards prevent mites that fall off bees from being able to re-enter the hive. Additionally, the introduction of honeybee strains that are more resistant to the mites shows promise. The biting mites also can spread viral and bacterial disease. Newer fumigant treatments do not show mite resistance.

African hive beetle: The African hive beetle (AHB) or small hive beetle, *Aethina tumida,* is a beehive parasite, endemic to sub-Saharan Africa. AHB can destroy colonies by causing damage to comb, stored honey and pollen. If a beetle infestation is sufficiently heavy, bees may abandon their hive. Great numbers of beetles have been associated with colony collapse disorder but should be considered a sign of a weak beehive for any reason. AHB can also be a pest of stored combs and honey in the comb awaiting extraction. Beetle larvae may tunnel through combs of honey, feeding and defecating, causing discoloration and fermentation of the honey, wrecking the comb, and ruining the honey. The adult beetle is dark brown to black and about 5 mm in length. Adults live up to 6 months and can be observed almost anywhere in a hive, although they are most often found on the inner cover, between frames, and on the rear portion of the bottom board. Female beetles lay their eggs in cracks or crevices.

The most effective control against AHB is a strong colony and a clean apiary. There are several traps currently on the market that can be placed in the hive; some are bait-able while others provide a safe, dark place to hide and trap them in non-toxic vegetable oil.

Tracheal mites: Acarapis woodi are internal parasites that live in the trachea or respiratory system of the honey-bee. The mites are capable of infecting all castes of the honey bee. Their presence in large numbers weakens the individual honeybee leading to less foraging, and a reduced lifespan. Tracheal mite infestation combined with other diseases lead to increased winter hive losses. Diagnosis can be made only microscopically buy dissecting out tracheoles of a dead honeybee. Infestation is fairly endemic in beehives but some strains of honeybees seem to be better able to tolerate low level infestations. There are several treatments such as thymol, formic acid or grease patties. The best defense is a strong colony with good food stores and if possible buying or selecting for tracheal mite resistance.

Summary

This very short paper is an introduction to the intricacies and interests of beekeeping, the honeybee and its health. There are many excellent books on the subject as well as a lot to be learned from beekeepers.

References

1. Caron, DM. *Honey Bee Biology and Beekeeping.* Kalamazoo, MI: Wicwas Press;2013.

2. Dadant & Sons. *The Hive and the Honey Bee.* Hamilton, IL: Dadant & Sons, Inc;1975.

3. Flottum K. *The New Starting Right With Bees.* AI Root Open Library;1997.

Honeybee Brood Diseases

S. Emmanuelle Knafo, DVM, Dipl ACZM

Session #2168

Affiliation: From the Department of Clinical Sciences, Cummings School of Veterinary Medicine at Tufts University, 200 Westboro Road, North Grafton, MA 01536, USA.

The most economically and medically significant honeybee diseases are diseases of the brood, as the health of the brood predicts the health of the colony. These diseases directly impact the growth and sustainability of the colony, and facilitate spread of pathogens due to the social structure of honeybees. The main brood diseases of importance to veterinarians treating honeybee hives are American foulbrood (AFB) and European foulbrood (EFB). American foulbrood, caused by *Paenibacillus larvae*, occurs when spores enter a colony and are fed to larvae by nurse bees. Spores can be carried by drifting workers, infected comb, contaminated tools, beekeepers, and hive robbing. There is currently no recommended treatment for AFB, and infected hives and equipment should be burned or otherwise destroyed by state apiary inspectors (www.apiaryinspectors.org). European foulbrood, caused by *Melissococcus plutonius*, is spread similarly to AFB, but treatment is possible. Improved hygienic beekeeping practices, nutritional support of the hive, and/or re-queening can help a hive recover. In more severely affected colonies, treatment with antibiotics may be indicated under the direction of a veterinarian. Other common brood diseases include chalkbrood (*Ascophaera apis*) and Sacbrood virus.

Honeybee Medications and Pharmacology

Gregory A. Lewbart, MS, VMD, Dipl ACZM, DipECZM (Zoo Health Management)

Session #2160

Affiliation: From the Department of Clinical Sciences, North Carolina State University, College of Veterinary Medicine, Raleigh, NC 27607, USA.

Abstract: The goal of these notes and the accompanying presentation are to introduce the reader/listener to what we know about honeybee medications and pharmacology, discuss some of the currently used drugs and the justification for doing so, and look ahead to what studies should or could be done in the near future. The emphasis of these notes is evidence based and references are included to support most statements and data.

Introduction

The honeybee (*Apis mellifera*) is one of the most economically important and widely studied animals on the planet. It is the only insect officially considered a food animal in the United States. In addition to producing honey, wax, and other products, many crops depend on honeybees for pollination. Collectively honeybees contribute about 15 billion dollars to the US economy each year[1] and over 25% of our calories come from plants pollinated by honeybees.[1,2] Globally, honeybees account for about 200 billion dollars in revenue, the majority coming from pollination.[2] Honey itself is a valuable commodity, and a single hive can produce 400 pounds of honey annually.[1] As veterinarians we need to be able to assist and support this critically vital component of agriculture, and when it comes to medications, do so in a responsible and legal manner.

Overview

Most of what we know about honeybee medications and pharmacology differs from larger invertebrates, and most vertebrates, in that we have almost no pharmacokinetic data based on individual animals. There are studies that examine drug residues, some of which are mentioned here, but most dosing is empirical at best. One of the most interesting behaviors of honeybees (of which there are many) is that of self-medication. There is strong support for the theory that some infected nurse and worker bees consume different types of honey based on their antimicrobial properties.[3-5] There also is a growing base of literature looking at the beneficial effects of honey and other honeybee products on treating humans, both therapeutically, and prophylactically.[6]

Although this is a talk about medications, one should keep in mind, that almost every medication has a downside or drawback. A recent study by Rayman et al. highlights the importance of the normal GI flora and they describe a dysbiosis in honeybees treated with antibiotics, leading to increased mortality, due to susceptibility to otherwise non-harmful pathogens.[7] In fact, the use of antibiotics for honeybees is illegal in Europe, though not in the United States.[7,8] For a relatively recent review of the topic of antimicrobials in honeybees see Reybroeck et al.[9]

Antimicrobials

Oxytetracycline

Oxytetracycline is used for treatment of American and European foulbrood in honeybees at a dose of 200 mg/colony every 4-5 days for three treatments.[10] Note, there is a lot of resistance now.[11,12] Those treating honeybees need to recognize they are considered food animals in the United States and proper Veterinary Feed Directive measures need to be followed.

Tylosin

Tylosin is used for treatment of American foulbrood in honeybees at a dose of 200 mg/colony every 4-5 days for 3 treatments.[9,10] Those treating honeybees need to recognize they are considered food animals in the United States and proper Veterinary Feed Directive measures need to be followed.

Parasiticides

Amitraz

Amitraz is used for control of mites (acariasis) in honeybees.[8,13] Use as directed but realize many factors are involved with a hive mite infestation.[13]

Formic acid

Formic acid is used for control of mites (acariasis) in honeybees.[8] Commercial packaging should be consulted prior to use.

Fumagillin

Fumagilllin is used for control of nosemosis (a microsporidian parasite) in honeybees.[8,18] Use as directed and consult packaging.[14-16]

Menthol

Menthol is used for control of mites (acariasis) in honeybees.[8,18] Commercial packaging should be consulted prior to use.

Thymol

Thymol is used for control of mites (acariasis) in honeybees.[8,17,18] Commercial packaging should be consulted prior to use.

References

1. Nolen RS. Federal directive brings veterinarians and beekeepers together. *J Am Vet Med Assoc News.* October 15, 2016.

2. Backues KA. The plight of the honey bee: what does this mean for one health? *Proc Annu Conf North Am Vet Comm.* 2017;1419-1420.

3. Erler S, Moritz RFA. Pharmacophagy and pharmacophory: mechanisms of self-medication and disease prevention in the honeybee colony (*Apis mellifera*). *Apidologie.* 2015;47:389-411.

4. Gherman BI, Dezmirean H, Dezmirean DS, et al. Defense strategies and self medication in honey bee (*Apis mellifera*). *Lucrari Stintifice-Seria Zootehnie.* 60:73-75.

5. Gherman BI, Denner A, Bobis O, et al. Pathogen-associated self-medication behavior in the honeybee *Apis mellifera. Behav Ecol Soc Biol.* 2014;68(11):1777-1784.

6. Miguel MG, Antunes MD, Faleiro ML. Honey as complementary medicine. *Integr Med Insights.* 2017;12: 1-15.

7. Raymann K, Shaffer Z, Moran NA. Antibiotic exposure perturbs the gut microbiota and elevates mortality in honeybees. *PLoS Biol.* 15(3):e2001861.doi:10.1371/journal.pbio.200861.

8. Vidal-Naquet N. Honeybee Veterinary Medicine; *Apis mellifera* L. Sheffield, UK: 5m Publishing;2016.

9. Reybroeck W, Daeseleire E, De Brabander HF, Herman L. Antimicrobials in beekeeping. *Vet Microbiol.* 2012;158:1-11.

10. Reynaldi F, Albo G, Avellaneda E, Rule R. Evaluation of two preparations of Tylosin administered in beehives for American foulbrood control. *Bulg J Vet Med.* 2016;doi: 10.15547/bjvm.1024.

11. Alippi AM, López AC, Reynaldi FJ, et al. Evidence for plasmid-mediated tetracycline resistance in *Paenibacillus* larvae, the causal agent of American foulbrood (AFB) disease in honeybees. *Vet Microbiol.* 2007;125(3-4):290-303.

12. Genersch E. American foulbrood in honeybees and its causative agent, *Paenibacillus* larvae. *J Invertebr Pathol.* 2010;103:Suppl 1:S10-9 doi: 10.1016/j.jip.2009.06.015.

13. Rinkevich FD, Danka RG, Healy KB. Influence of Varroa mite (*Varroa destructor*) management practices on insecticide sensitivity in the honey bee (*Apis mellifera*). *Insects.* 2017;8(1):9.

14. Bailey L. Effect of fumagillin upon *Nosema apis* (Zander). *Nature.* 1953;171:212-213.

15. Giacobino A, Rivero R, Molineri AI, et al. Fumagillin control of *Nosema ceranae* (Microsporidia: Nosematidae) infection in honey bee (Hymenoptera: Apidae) colonies in Argentina. *Vet Ital.* 2016;52(2):145-151.

16. Higes M, Nozal M, Alvaro A, et al. The stability and effectiveness of fumagillin in controlling *Nosema ceranae* (Microsporidia) infection in honey bees (*Apis mellifera*) under laboratory and field conditions. *Apidologie.* 2011;42:364-377.

17. Carayon JL, Téné N, Bonnafé E, et al. Thymol as an alternative to pesticides: persistence and effects of Apilife Var on the phototactic behavior of the honeybee *Apis mellifera*. *Environ Sci Pollut Res Int.* 2014;21:4934. doi:10.1007/s11356-013-2143-6

18. Vidal-Naquet N. Honeybees. In: Lewbart GA, ed. *Invertebrate Medicine.* 2nd ed. Ames, IA: Wiley-Blackwell; 2012:285-321.

Integrated Pest Management in Beekeeping

Kay A. Backues, DVM, Dipl ACZM

Session #2162

Affiliation: From the Tulsa Zoo, 6421 E 36th St N, Tulsa, OK 74115, USA.

Abstract: The European honeybee, *Apis mellifera* is an agriculturally important animal. The human management of bees, apiculture has been practiced for 1000s of years but only in the last 50 or so years has become a large scale global industry. The demands of industrial scale agriculture require that a pollinator be present in artificially high numbers at crucial times of the year to pollinate food crops. The European honeybee is the most widely used and useful human managed insect for this purpose, contributing most of the estimated $14 billion dollars in pollination services required for the US food supply. Globalization has led to a number of diseases, and pests that have been introduced to the honeybee. The last 20 years have seen dramatic bee die-offs and an industry response to try to treat the honeybee for various maladies. Similar to pathogens treated in more common veterinary and human patients, honeybee pathogens have shown the rapid development of resistance to first-line treatments. The treatments themselves have led to unforeseen consequence for the bee and the colony such as reduced fertility in queen bees. The principles of integrated pest management have many avenues for application in honeybee hives to lessen the effects of chemicals and keep these important agricultural and food producing animals healthy.

Integrated Pest Management

Integrated pest management (IPM) is an ecosystem-based strategy that focuses on long-term prevention of pests or their damage through a combination of techniques such as biological control, habitat manipulation, modification of cultural practices, and use of resistant varieties. The principles of IPM to lessen, prevent or treat diseases, and parasites of the European honeybee requires a thorough understanding of the lifecycle of the hive, the disease or parasite involved and the time of the year for the hive. IPM should strive to use less chemicals like antibiotics and acarides but it does not mean that such treatments should never used. Antibiotics and acarides labeled for beehives should be avoided for prophylactic use when IPM strategies can achieve the same goal.

Honeybee Bacterial Diseases

Two important bacterial diseases affect the honeybee. They are described below.

American foulbrood (AFB)

This is the most important bacterial disease that affects bee larvae. AFB is caused by spore-forming *Paenibacillus larvae* ssp. *larvae* (formerly *Bacillus larvae)*. The spores can be brought to the hive by bees that drift in from a different hive or bees from the hive that have robbed a sick hive. The spores contaminate food that is fed to the larval bees. House bees remove dead brood which is contaminated with the bacterial spores and thus spread the spores all over the hive. The spores cannot be removed from the hive and are essentially immortal, remaining viable for decades. AFB infected hives weaken and die and act as a source of infection for other hives.

Treatment will only temporarily mask signs of the disease as the spores persist in the hive indefinitely. Antibiotics labeled for use in the beehive directed at European foulbrood and AFB are available but their use is controversial. Prophylactic treatment to prevent AFB and EFB are no longer recommended due to the risk of developing resistance. A confirmed infected AFB hive should be sealed and burned so that no bees can spread the disease and none of the infected hive parts can be reused. In many states AFB is a reportable disease to the state agricultural division and the state may require the burning affected hives. A few state agricultural offices offer irradiation of contaminated hive equipment, hives, frames, and woodenware. The bees should be destroyed to prevent the spread of the disease.

European foulbrood (EFB)

Bee larvae become infected with European foulbrood when they consume brood food that contains the bacteria, *Melissococcus plutonius*. The bacteria reproduce in the gut and compete for the food that is fed by nurse bees to the larva. Depending on the level of infection and amount of available food, the infected larva will either survive or die. The larvae that survive go on to defecate and pupate, which leaves bacteria on the combs that can be infective for years, even though this bacteria does not produce spores. Treatment with antibiotics will help stop the disease but a strong hive and queen are better preventatives.

Fungal Diseases

There are several fungal diseases of brood such as chalk brood and stone brood caused by the fungi *Ascosphaera apis* and *Aspergillus* sp. respectively. These diseases maybe seen sporadically in late spring or when atmosphere in the hive is not optimal. Again a strong laying queen is the best preventative as these fungal diseases are associated with not enough workers to keep brood warm or ventilate hive appropriately.

IPM Strategies for Treatment of Bacterial and Some Fungal Diseases

IPM strategy #1. The first and often most effective treatment for brood diseases in beehives is having a strong laying queen. Queens are most fecund their first 1-3 years of life and most beekeepers replace queens every 1-2 years. But a honeybee queen should be replaced anytime her performance is not deemed adequate. A strong laying queen makes more workers to keep hygiene in the hive optimal, and moderate temperatures which reduces stress to larvae. Large numbers of nurse bees can provide more food for larvae and keep the larvae warm during spring when sudden temperature drops can happen. All diseases listed will benefit from a young strong laying queen.

IPM strategy #2. Routine brood frame replacement. To prevent the buildup of fungal and bacteria and bacterial spores in a hive old brood frames should be replaced each year. It is often not feasible to replace all the precious brood nest comb every year but letting the frames be used indefinitely allows for the abnormal buildup of pathogens. Strive to replace 10-20% of all frames in a hive each year, this would be done by removing the oldest frames in the spring and putting in fresh brood frame foundation for young bees to draw new wax on. The old frames should be bagged up and discarded. They should not be left where bees have access to them as they will investigate these combs, rob any honey present and bring the pathogens back with them to the hive.

IPM Strategy #3. A screened or ventilated bottom board allows for better ventilation throughout the hive and also helps with Varroa mite infestation. Good ventilation to prevent condensation of moisture in winter and overheating in summer. The ventilated bottom boards help ensure air flow and provide a way help reduce Varroa mites, which if they fall off of bees cannot climb back up from the ground and enter the hive.

Antibiotics can be used for treating EFB when properly diagnosed and with a VFD. Treatment for AFB is controversial due to the immortal nature of its spores. The state agriculture dept should be contacted for all AFB diagnoses to ensure proper compliance with state regulations.

Parasitic Diseases

Mites

Varroa mite: Varroa destructor appeared in European honeybees in the US in the 1990s and led to a significant die off of hives. The Asian honeybee, *Apis cerena*, is the natural host the Varroa mite and is resistant to significant disease from it, but Varroa is considered one of the most significant diseases of European honeybee colonies. Without treatment hives can weaken and die within 1-3 years. Heavy infestations lead to a weakened hive late in the season that likely will not survive the winter. Spread of deformed wing virus is a telltale sign of heavy infestation of Varroa. Transmission of Varroa infestation is likely from infected bees drifting to other hives, or weak, infected hives being robbed by other bees: they steal the honey but also bring the mite back to their own hives.

Tracheal mite: Acarapis woodi are internal parasites that live in the trachea or respiratory system of the honeybee. The mites are capable of infecting all castes of the honey bee. Their presence in large numbers weakens the individual honeybee leading to less foraging, and a reduced lifespan. Tracheal mite infestation combined with other diseases lead to increased winter hive losses. Diagnosis can be made only microscopically buy dissecting out the respiratory tubes within the bees thorax, the tracheoles of a dead honeybee. Infestation is fairly endemic in beehives but some strains of honeybees seem to be better able to tolerate low level infestations.

IPM Strategies for Mites

A strong fertile queen is your first important hedge against disease from Varroa and tracheal mites. Weak hives are easily overwhelmed by mites, Varroa or tracheal infestation. Weak hives may not survive the winter if heavily infested with tracheal mites. Strong hives can tolerate these parasites at low levels.

- Select a strain of honeybee with known Varroa tolerance, the tolerance comes from high hygienic behavior. The trait is known as selective mite resistance, SMR. House bees of this honeybee trait are highly motivated to clean out cells, detect and take out infected larva and disrupt life cycle of Varroa mite. Hygienic bees also practice increased grooming of their hive mates and may chew and damage Varroa mites.
- There are now many strains of honeybees that are tracheal mites resistant.
- There are now strains of honeybees that are showing good Varroa mite resistance.
- Ventilated bottom boards, so if Varroa mites fall off they cannot get back to beehive.
- Selected methods to help Varroa mites fall off such as powder sugar dusting, etc
- Sugar and shortening patties fed to bees in late summer and fall reduce tracheal mites ability to infest newly hatched bees.
- Chemical treaments and acarides also control tracheal mites but are used primarily for Varroa control.

- Chemical treatments that inhibit mites such as Thymol and Formic acid. These exploit weakness of mites, but mites cannot become resistant.

- A menthol packet placed over brood nest in winter prevents tracheal mite disease via desiccating the mites.

- Using a labelled acarides only after doing mite drop counts and determining a threshold level of mites has been reached and acaride is only way to control infestation. Using these acarides prophylactically every year has led to resistance and lowered fecundity of queens raised in hives with their chemical residue.

- Labeled for Varroa mite: Apistan and Check Mite are brand names.

All chemical treatments must be used only when honey collection is not ongoing in hives. Typically such treatments are performed in late summer after honey supers have been removed. For all the Acarides and other treatments such a Thymol a withdrawal time from use before supers can be applied is on the label. Honey is a human food product and label withdrawal times must be strictly followed!

Insects

The African hive beetle (AHB) or small hive beetle, *Aethina tumida,* is a *beehive* parasite, endemic to sub-Saharan Africa. AHB can destroy colonies by causing damage to comb, stored honey and pollen. If a beetle infestation is sufficiently heavy, bees may abandon their hive. Great numbers of beetles have been associated with colony collapse disorder but should be considered a sign of a weak beehive for any reason. AHB can also be a pest of stored combs and honey in the comb awaiting extraction. Beetle larvae may tunnel through combs of honey, feeding and defecating, causing discoloration and fermentation of the honey, wrecking the comb, and ruining the honey. The adult beetle is dark brown to black and about 5 mm in length. Adults live up to 6 months and can be observed almost anywhere in a hive, although they are most often found on the inner cover, between frames, and on the rear portion of the bottom board. Female beetles lay their eggs in cracks or crevices.

The most effective control against AHB is a strong colony and a clean apiary. There are several traps currently on the market that can be placed in the hive; some are bait-able while others provide a safe, dark place to hide and trap them in non-toxic vegetable oil.

Summary

Integrated pest management strategies can be used by the knowledgeable beekeeper to reduce chemical usage in their hives. A thorough understanding of what can be done takes some practice and extra steps to keep beehives healthy. There are many excellent books on the subject as well as a lot to be learned from local beekeepers.

References

1. Caron, DM. *Honey Bee Biology and Beekeeping.* Kalamazoo, MI: Wicwas Press;2013.

2. Dadant & Sons. *The Hive and the Honey Bee.* Hamilton, IL: Dadant & Sons, Inc;1975.

3. Flottum K. *The New Starting Right With Bees.* AI Root Open Library;1997.

Colony Collapse Disorder

S. Emmanuelle Knafo, DVM, Dipl ACZM

Session #2169

Affiliation: From the Department of Clinical Sciences, Cummings School of Veterinary Medicine at Tufts University, 200 Westboro Road, North Grafton, MA 01536, USA.

Colony collapse disorder (CCD) is a term used to describe a phenomenon in which hives die due to rapid loss of adult worker bees, leaving behind the queen, food, nurse bees and brood. Without the worker bees to bring nectar and pollen to the hive, the colony becomes weak and dies (collapses). Signs of hives affected by CCD include weak or dead colonies with excess brood in relation to number of adults present, and the lack of dead workers within and around the hive. Typically, there are no signs of an infectious etiology, and no significant *Varroa* mite or *Nosema* loads. Colony collapse disorder causes over-wintering losses that can have massive economic effects on beekeepers and impact honey bee populations. There is currently no clear cause of CCD, but it is likely a multifactorial phenomenon. Hypothesized causes include the widespread use of pesticides in modern agriculture, particularly neonicotinoids, the rise of monoculture crops coupled with a decline in meadows and wildflowers as nutritional resources for bees, the introduction of varroa mites, intensive honey bee management practices and associated stress from hive medications and transport for pollination services.

The Use of Honey in Veterinary Medicine

Susan Horton, DVM

Session #2156

Affiliation: From Chicago Exotics Animal Hospital, 3757 W Dempster St, Skokie, IL 60056, USA.

Abstract: Medicinal honey can be used for treating wounds. It can be used to treat burns, chronic or acute ulcers, abscesses, and deep cavity wounds. Honey is one of those ancient medicines that has been used to treat all sorts of ailments in the ancient human record. It is a therapy that has enjoyed new popularity in veterinary medicine. It is very helpful in wound management.

Wound-healing Characteristics of Honey

Honey has features that reduce bacterial populations in the wound. It has high osmolality (hyperosmotic). This means that it attracts any free water in the wound making it unavailable to the microorganisms. Hydrogen peroxide is produced through an enzymatic reaction when honey is diluted by wound exudates. This is released slowly to provide antibacterial activity. Some honeys have phytochemicals (Manuka Leptospermum honey) that are antibacterial. This type of honey in particular stimulates wound healing by controlled release of cytokines.

Honey is most often used to treat wounds such as abscesses, burn ulcers, and chronic non-healing dermal ulcers. It is important that a sterilized, laboratory tested honey for medicinal purposes be used. Many different dressings are available.

Comparing honey with other wound healing products, consider efficacy, reduced potential for antimicrobial resistance, safety within and around the wound, ease of application, availability and cost. Considering the common antibiotic resistance encountered routinely in abscess and wound management, honey provides an alternative therapy that does promote microbial resistance.

How to Use Honey in Wounds

The amount of honey depends on how much exudate is present in the wound. Highly exudative wounds will require frequent changes. Things like abscesses or deep cavity wounds need more honey to penetrate deep into the tissues before applying the honey dressing. The honey will promote debridement of necrotic tissue, anti-inflammatory action, and speed granulation bed formation.

Medihoney (Derma Sciences, Ontario, Canada) makes a variety of products for wound management. Most recently I have been trying the calcium Alginate dressing with active Leptospermum honey as a packing in deep cavity abscesses. This works with larger spaces. It tiny, tight spaces, it becomes a melty mess.

The hydrogel non-adhesive squares work well for burns and large areas of secondary wound healing. They do come with adhesive, but I find the adhesive difficult to work with. Apply the gel square to the clean surface of the burn/tissue, then bandage in place. Change daily. The gel can also be applied directly to the tissue before the square. This does seem to be less painful as far as bandages that need to be changed daily go.

The gel itself is handy for prolapses. It is effective in reducing edema as well as helping clean up the secondary bacterial and necrotic issues. Allow the tissues to soak for 1-2 hours, rinse and replace. Honey can be instilled after reduction. It will be expelled naturally.

References

1. Bang LM, Bunting C, Molan P. Peroxide production in honey and its implications. *J Altern Complement Med.* 2003;9:267-273.

2. Castiglioni S, Stefano M, Astolfi P, Carloni P. Chemometric approach to the analysis of antioxidant properties and colour of typical Italian monofloral honeys. *Int J Food Sci Technol.* 2017;52:1138-1146.

3. Mathews K, Binnington A. Management of wounds using honey. *Compend Contin Educ Pract Vet.* 2002;24 (1):53-60.

4. Molan PC. Re-introducing Honey in the Management of Wounds and Ulcers-Theory and Practice. *Ostomy Wound Manage.* 2002;48(11):28-40.

5. Molan PC. The evidence supporting the use of honey as a wound dressing. *Int J Low Extrem Wounds.* 2006;5:40-54.

6. Pask J. Jessica Pask. Antibacterial Efficacy of Honeys from Southwestern Ontario. Master's thesis. University of Guelph, Ontario, Canada;2013.

7. Subrahmanyam, M. A prospective randomized clinical and histological study of superficial burn wound healing with honey and silver sulfadiazine. *Burns.* 1998;24(2):157-161.

8. MediHoney website. Availabe at: http://www.dermasciences.com/medihoney-products Accessed June 30, 2017.

The Veterinary Feed Directive Rules for Veterinarians: A Practical Guide for the Practicing Veterinarian

Lance Roasa, DVM, JD

Session #2163

Affiliation: From Flatwater Veterinary Group PC, The Roasa Law Group, Drip Learning Technologies, LLC and University of Nebraska-Lincoln School of Veterinary Medicine and Biomedical Sciences.

Abstract: The transition of commonly used antibiotics away from production based -label claims to Veterinary Feed Directive (VFD) status is one of the largest regulatory changes to food animal veterinary practice in decades. This purpose of this document to aggregate information, aid veterinarians and their practices in making the transition, and provide practical solutions to questions created by the new VFD rules.

Note: The information contained in this document is not intended to create an attorney client relationship and should not replace legal counsel. These materials are for information purposes only and should not be taken as legal advice. The reader is encouraged to seek an attorney for advice on specific legal problems.

Reprinted with permission of Lance Roasa, DVM, JD and presented by Micah Kohles, DVM, MPA.

Why Are These Changes Occurring?

The VFD rules are designed to give veterinary oversight into medically important antibiotics.

The Food and Drug Administration (FDA) is responsible for overseeing the marketing and use of drugs. Historically, two classifications of drugs, over the counter (OTC) and prescription, were available to drug manufacturers applying to market and sell drugs in the United States. Once the FDA approves a drug through the appropriate classification, the drug's label places requirements and restrictions for its use and that label controls the marketing and use of that drug. The designator "medically-important antibiotics" does not apply to all antibiotics, just those that are deemed useful to human medicine and are at-risk to antibiotic resistance.[1]

In 1996 Congress created a third classification by which drug manufacturers could bring drugs to market, the VFD.[2] Few drug manufacturers utilized the VFD pathway over the next 20 years[3]; however, with a series of guidance documents, the FDA announced its intention to request that manufacturers voluntarily transition OTC labeled drugs to the VFD label claim. This cumulated in April 2015 with a final rule promulgated by the FDA that updated the requirements for using VFD drugs and the respective responsibilities of veterinarians, producers and feed distributors.[4]

Recently, the FDA has published several Guidance For Industry documents designed to aid the transition process.[5] These guidance documents are written in plain language and help apply the new rules to actual scenarios. The guidance documents leave some information gaps in the real world practice of food-animal medicine. This document will summarize the FDA's publications and supplement practical information where needed.

The FDA and World Health Organization have determined that medical evidence links antibiotic resistance in humans to antibiotic use in food producing animals.

Understanding the reasoning behind the rules may help a veterinarian's appreciation of the changes and assist with compliance. The reasoning behind the current transitioning of drug labeling to VFD status is the concern and evidence regarding antibiotic resistance in humans. As early as 1970, the FDA began to examine the safety and efficacy of penicillin and tetracycline used in animal feeds.[6] In 1977 the FDA proposed public hearings on the matter.[7] Shortly thereafter, Congress directed studies regarding the possibility of human antibiotic resistance to begin.[8] The FDA delayed the hearings, pending the results of the research. Ultimately, the FDA never held the hearings regarding antibiotic use in feeds and nearly 40 years later, advocacy groups filed lawsuits demanding that the FDA hold the required hearings.[9]

In the 45 years since researchers have begun to examine the issue of human antibiotic resistance stemming from sub-therapeutic use in food animal feeds, the FDA has used task forces, expert panels, WHO reports, Government Accountability Office reports, the National Research Council reports, Institute of Medicine reports, and peer reviewed scientific literature to provide a evidentiary framework for the rule.[10] Veterinarians interested in the research and reasoning behind the conclusion are encouraged to read Guidance for Industry Document #209, where the FDA outlines its thinking and cites peer-reviewed literature that the rule is based upon.[11] The scientific debate concerning the link between antibiotics in animal feeds and antibiotic resistance bacteria is human medicine is ongoing.[12] The issue is complex and multifaceted. Normal dirt flora, metals such as Zinc, and non-antibiotics such as salicylate all potentially contribute to human antibiotic resistance.[13] Additionally, human use of antibiotics, veterinary use in non-food species, and cross-resistance even across antibiotic classes create continued resistance despite complete antibiotic bans.[14]

Although the research is ongoing, and the scientific debate ensues, the FDA has promulgated rules that will carry the force of law beginning January. 1, 2017. A number of commenters have written that the VFD rules don't go far enough and that they leave large loopholes where antibiotics will continue to be fed for preventative but non-production reasons.[15] Regardless of the debate and the unsettled science behind the rules, veterinarians have the responsibility to know the rules, and abide by their requirements. Veterinarians will be in a position of oversight and education for the entire VFD process. This oversight component has thrust veterinarians into a position where they stand between feed distributors and producers who use antibiotics in their animals. This position will cause inconveniences and even disputes between the parties; however, this is designed to provide oversight and involvement by the veterinarian. This involvement could lead to new means to show how a veterinarian's input could benefit the producer, not hinder them. Veterinarians who grasp this opportunity should see growth in other practice areas where they will regularly have the opportunity to display their knowledge; however, this increased responsibility has the potential to bring additional liability to veterinarians and their practices, mitigating that risk will be discussed fully below.

What are the Practical Aspects of the VFD Process that a Veterinarian Needs To Know?

The process and specific requirements of the VFD process are well characterized in the reference materials provided.[16] Please refer to this practice summary sheet for the details regarding writing, issuing and storing the VFD.[17] Below are issues that are not addressed by either the FDA or other reference materials regarding the VFDs, but should nonetheless be incorporated into the veterinarian's plan for VFD compliance.

A veterinary client patient relationship is needed before issuing a VFD.

The VFD Rules require that a Veterinary Client Patient Relationship (VCPR) be established before issuing a VFD.[18] Many states have specific requirements for establishing a VCPR in their practice acts or applicable regulations.[19] For other states that don't address the VFD process directly, a federal regulation provides a VCPR requirement for VFD drugs.[20] Regardless, the underlying theme is that the "issuing veterinarian is assuming the responsibility for making medical judgments regarding the health of the animal(s)" and "the client has agreed to follow the instructions of the veterinarian."[21]

The federal VCPR requirement continues with specifics as to how the veterinarian is to assume the responsibility. The key language in the federal VCPR that a veterinarian must be aware of is "there is sufficient knowledge of the animal"… "by virtue of examination and/or medically appropriate and timely visits to the premises where the animal(s) are kept." In the daily practice of food animal and mixed animal practice it is practically impossible to perform physical examinations on every animal that will be fed medicated feed. Concurrently, in many veterinary practices, an ongoing client relationship is established with many clients through visits to their facilities. The VCPR regulatory language leaves clinical discretion to the veterinarian to determine the amount of "sufficient knowledge" of the facility to issue a VFD.[22]

Another potentially impactful requirement of the VCPR is the necessity for a valid license in the state where the animals are fed.[23] This mean that veterinarians overseeing animals on feed in numerous states, even if those animals are owned by the same entity, must maintain licensure in all states where the medicated feed or water is fed or consumed.

Lastly, the federal VCPR requirement states that the issuing veterinarian be "readily available for follow up in case of adverse reactions or failure of the regimen of therapy."

The veterinarian should regularly visit the facility to maintain a valid VCPR.

Neither the rule creating the VFD, the federal VCPR requirement, nor the FDA's publications state a hard and fast rule for the frequency of visits that a veterinarian needs to retain the "sufficient knowledge" to issue VFDs; however, the maximum time before expiration of the VFD is 6 months from the date of issuance.[24] A proposed guideline for maintenance of a VCPR could state that maximal time between facility visits is 6 months, or the applicable expiration date.

When on the facility, the veterinarian can use this opportunity for education, maximizing return and showing value for veterinary advice and input to the producer. While on the facility, the veterinarian could check animal husbandry items such as housing conditions, animal handling, nutrition, vaccination protocols and other routine items. Properly addressed, these areas can provide increased returns through improved animal welfare. A knowledgeable veterinarian in these areas could provide great value for the producer and could potentially offset the cost of the VFD. Next, for concentrated animal feeding operations, amount of pen space, manure disposal practices, water access and runoff could all contribute to antibiotic residues in the environment. Some published research papers discuss the possibility that antibiotics and their metabolites, excreted via urinary and fecal routes, as a potential pathway for building antibiotic resistance.[25] The veterinarian could educate and instruct the producer regarding these practices, which could have significant impact on future actions and potential producer liability.

At the very least, making and documenting recommendations for facilities with deficient welfare practices and environmental issues is an important part of the onsite veterinary practice; however, it is unaddressed in a cursory glance of the VFD rules. An understanding of the current legal environment facing CAFOs may help veterinarians reduce liability for their production animal clients.

The client is the person responsible for feeding and caring for the animals.

Although the VCPR requires communication between the veterinarian and the client, determining the client in many food animal situations may be confusing. Animals may be owned by the facility, a distant producer with retained ownership, or a corporate entity. In these cases, it is impractical for a veterinarian to contact and form a relationship with those owners. For simplicity, the FDA has determined that the client may be the actual owner or the caretaker.[26] Consequently, the person or entities paying the veterinary bills or the legal owner of the animals may not be the "client" for purposes of the VFD.

The use of production label claims will be phased out and only treatment, control or prevention claims will be allowed.

A significant impact on the use of many antibiotics is the transition from production based label claims, such as "feed efficiency or "increased weight gain" to therapeutic label claims. Potential therapeutic label claims include "prevention" "treatment" and "control" of a particular disease. For example, the label for Auromycin® (chlortetracycline), a commonly used feed additive in beef cattle, states "for the treatment of bacterial enteritis and pneumonia."[27] Other labeled uses for Auromycin® include "As an aid in control of active infection of anaplasmosis caused by Anaplasma marginale susceptible to chlortetracycline."[28] The claims are separate and distinct from the previous label claims, which stated, "increased rate of weight gain and improved feed efficiency."[29]

The FDA's implementation strategy in the transition was that of voluntary withdrawal of the production claims and reapplication for prevention and/or treatment claim by the sponsor (manufacturer). As of June 30, 2016 all sponsors affected by the FDA guidance for industry documents have agreed in writing to voluntarily withdraw their production or feed efficiency label claims.[30] By the date of implementation of the VFD rules on January 1, 2017 all labels for medically important antibiotics in feed additives and medicated feeds should be transitioned.

The veterinarian must know and follow the label on VFD drugs.

The particular label claims of various VFD drugs will become very important for veterinarians to know and use. Extralabel use of drugs in food-producing animals is never allowed and is often the subject of FDA action against veterinarians.[31] The label will be the source of information regarding the indications for use, the maximal duration of feeding, any potential medicated combinations that could be fed, and any withdrawal times.[32] The veterinarian must write the VFD with pertinent labeling requirements in mind. The veterinarian will be viewed as the expert on the label claims when providing oversight to the antibiotic feeding process. Further, the veterinarian will often be involved with the explanation and edification of the label to the producer because a diagnosis is required and thus medical information is involved. A veterinarian that complies with label instructions will greatly reduce their liability risk. In summary, the veterinarian needs to thoroughly understand the labels for VFD drugs that are commonly used in their practice area, by their clients, producers, and feed distributers.

The VFD drugs' labels have specific species and indications on the label. The veterinarian issuing the VFD cannot deviate from the label regardless of whether the animal is intended for food production. The prohibition on extra-label use forbids use of the VFD drug in minor species not listed on the label, for major species in non-food production use, and for conditions not listed on the label.[33]

The medical record should reflect the issuance of a VFD.

Another large impact on veterinarians is the recordkeeping requirements brought about by the new rules. The FDA's guidance is very detailed on both the information that must be recorded, and the method of recordkeeping of the actual VFD[34]; however, the FDA's publications are silent as to the medical recordkeeping requirements of the veterinarian and practice. Most states have statutes or regulations that require detailed written medical records to be kept by the veterinarian.[35] Practice to the standard of care almost always requires that a medical record be produced for each VCPR. The medical record is the veterinarian's main defense in later civil actions and/or state licensure actions. Additionally, the AVMA's Principles of Veterinary Medical Ethics, or the ethical requirements of the profession, state that records are an integral part of veterinary care.[36] Lastly, but of extreme importance, practices with good medical records tend to have more efficient communications among veterinarians and paraprofessionals, provide clearer communication to clients, provide better patient outcomes, and tend to have more value at practice transition than practices without adequate medical records.[37]

To issue a VFD a veterinarian must have "sufficient knowledge" of the production facility through "physical examination" or "visit of the facility."[38] How the veterinarian gained the knowledge to issue the VFD should be documented in the medical record.

The medical record should state a diagnosis for the herd that is consistent with label claims on the VFD drug.[39] Practically, it is impossible to definitely diagnosis every animal in a herd health setting. Some label claims will be made for preventative uses. Moreover, some diagnoses can only be made on post mortem examination. The FDA has been clear that a presumptive diagnosis, made with the veterinarian's clinical judgment is sufficient, but makes no requirement that the diagnosis be documented on the VFD form.[40]

The mechanism for enforcing the validity of the presumptive diagnosis, along with the veterinarian's clinical judgment, are highly necessary to meet the standard of care in a civil or state board action.[41] The medical record should support both the presumptive diagnosis and the veterinarian's clinical reasoning for issuing a VFD. Writing the record with the goal of indicating clearly how the veterinarian gained "sufficient knowledge" and how they arrived at a "presumptive diagnosis" for the label claim on the VFD drug will significantly decrease civil and board action induced liability to the veterinarian.

The medical record can incorporate the VFD and the information included within it by reference, so there is no need to repeat redundant information. Information not found on the VFD, but which is still relevant and important to include within the medical record, including the physical exam or facility findings, pertinent information regarding the history, husbandry and intended use of the animals, the presumptive or definitive diagnosis and any other relevant client communication regarding the animals. The Subjective, Objective, Assessment, Plan (SOAP) manner of writing medical records could be slightly modified to capture all pertinent aspects of the document, but by using a herd health approach instead of a individual patient approach.

The identification and location of the animals to be fed the VFD will pose logistical concerns for the veterinarian. Many veterinary hospitals do not currently use software packages or paper filing systems that allow different parties for the client and owner.[42] Data input fields are often not available to allow the practice to capture different locations. For purposes of filing and searchability, the practice should standardize the manner of data entry to input the herd health and VFD data. A medical recordkeeping system of standardized nomenclature of herds receiving VFD drugs could be developed by the practice. Since the herds don't have given names like individual animals, the name of the patient in the medical record is created by the practice personnel. For example, a herd health system of recordkeeping could be analogous to the animal(s) identification on the VFD. The name field could state "50 Angus heifers received into pen 11C on 4-1-17," then pertinent medical information could be entered. This name could also abbreviated with a standardized practice naming convention to "50 Ang H 11C 4.1.17."

Practices currently utilizing handwritten records should strongly consider converting to electronic medical records (eMRs) for animals being fed VFD drugs, due to the recordkeeping requirements for large producers and practices issuing large numbers of VFDs. Cloud-based recordkeeping systems that are mobile and automatically updated via internet connection make eMR systems appealing to ambulatory practitioners. Tablet-based systems are becoming more functional and these technologies could become a labor saving technology for mobile food animal veterinarians issuing even moderate numbers of VFDs.

Electronic transmission and storage of the VFD is allowed.

Another time and labor saving technology that will improve the efficiency of the VFD process for veterinarians and their offices is the electronic transmission of the VFD. The VFD rules clearly state that the veterinarian is responsible for delivery to the feed mill or distributer.[43] The entire form can be electronic; even the signature of the issuing veterinarian.[44] Paradoxically, VFDs that are emailed are not considered "electronic VFDs" but are considered to be transmitted electronically.[45] True electronic VFDs are generated and transmitted via the internet generally by a third party service that is neither the veterinarian nor the feed distributer.[46] The most simplistic version of the transmission and storage of a written VFD is on an internet-based email or server system.

The VFD must be retained by the veterinarian for 2 years, either in its electronic or paper form. The VFD rules do not specify the location or security of the VFD. The most simplistic and practical version of transmission and storage would be an electronic tablet with a fillable VFD form.[47] The form could then be signed by the veterinarian.[48] An email of the VFD could be emailed to the producer and the distributer and carbon copied to the veterinarian before leaving the facility. The carbon copy (cc) version could then be attached to the producer's electronic medical record to be stored electronically and in a folder within the email service.

The VFD must contain a physical location.

The VFD rules require that the street address, GPS coordinates, or physical characteristics such as pen, barn, tank or other distinguishing characteristics be listed on the VFD. The FDA states that it understands that animals may come and go from lots of animals and animals may move to other locations. Therefore, it is acceptable to list multiple locations on the VFD.[49] This exception is caveated by the requirement that the veterinarian needs to be licensed in each state and the feed for one VFD must come from a single manufacturer.

There is no standardized VFD form, but the AVMA, the FDA, and drug manufacturers have provided examples.

The veterinarian can choose to create their own paper VFD form as long as the regulatory requirements are met.[50] The FDA has distributed an example that can be used or modified.[51]

The entire VFD form creating process could be completed online, with electronic transmission and storage online. The veterinarian's signature could be electronic, eliminating the need for printing and scanning of the document.[52] For efficiency, time and cost savings, and reduction of errors, it is suggested that veterinary practices streamline their process with the use of technology. Practically, the less time that a veterinarian spends writing and transmitting the VFD, the more efficient and productive they can be in other areas of practice. A more efficient process will aid the veterinary practice in maintaining profitability while complying with the VFD rules.

The veterinarian can charge the client for forming the VCPR and issuing the VFD, but they will need to become as efficient as possible.

Food animal veterinary practice is labor intensive, time consuming, and producers are cost sensitive. Practically, it will be difficult for a veterinarian to charge appropriately for the time and travel costs to visit and examine animals or inspect facilities. For a veterinarian to establish a presumptive or definitive diagnosis will take time that was not previously required when medically important antibiotics could be obtained over the counter. Additional communication with the client will be needed to establish the approximate number of animals, the intended use of the animals, and the expiration of the VFD. Client education will be a major component of these changes. The practicing veterinarian will have a large initial time outlay in order to gain the knowledge to issue VFDs that are compliant for labeling, combination drug usage, withdrawal times and cautionary statements. Lastly, the transmission and storage of the VFD and the increased medical recordkeeping implications will be time-consuming for the veterinarian and paraprofessionals. Producers not currently budgeting increased veterinary fees for the increased veterinary oversight will be price sensitive and at times could become frustrated with the veterinary practice for charging for the veterinarian's time. Still though, it will be difficult to properly charge for the veterinarian's time, unless the practice is efficient and streamlined in the research, writing, transmission and storage of the VFD.

Communicating with producers every time a facility visit is made about the possibility of a future VFD need could reduce single-purpose visits made to issue a VFD. Also, coordinating and combining visits to distinct production facilities in rural areas could decrease drive time and lower costs of implementation. Training para-professionals to obtain standard history, perform data entry, and provide common client education could limit veterinarian input time.

Practices that are in heavy food animal production regions may consider hosting client education workshops or events. Handing out brochures, client information literature and directing clients to online information may provide additional time savings.[53] Veterinarians will certainly want to work with local feed mills (distributers), retailers, and large production facilities to streamline and standardize their communications and VFD transmission.

A producer can mix medicated feed on the farm and a veterinary hospital can sell medicated articles or feed.

Many animal producers mix feed on their facility for immediate feeding, using bulk-quantity medicated crumbles or mix.[54] These producers currently buy the drug in bulk from a retailer/distributer. After implementation of the VFD rule, the producer will still be able to mix medicated feed as long as the labeling and VFD instructions are met.

The veterinary practice can sell medicated Type A (medicated articles) or Type B (mineral/supplement mixes) directly to producers for use on their facility. To do this, the veterinary practice would need to submit a one-time letter of distribution to the FDA and also ensure that each producer has valid VFD before purchase.[55] If the veterinary practice were to sell to other distributors or entities then mixing or selling feeds for other producers, then the veterinary practice would need to keep a "acknowledgement letter" from the buyer of the feed.[56]

To purchase the VFD drug, the retailer will need to maintain a letter of distribution on file from the producer and a valid VFD for the drug to be purchased.[57]

In conclusion, the new VFD rules represent some of the largest regulatory changes to food animal veterinary practice in decades. Veterinarians practicing in production and mixed animal settings need to understand the rules, guidance documents, and VFD drug labels. Of particular importance for veterinarians is determining how to establish a valid VCPR with the sufficient knowledge needed to write the VFD. Further, veterinarians should be diligent in creating thorough medical records that document the presumptive diagnosis and how the veterinarian found a need for a VFD drug. Veterinarians could take advantage of increased oversight to provide value for their services to clients. Lastly, veterinarians and their practice's offices should work to become as efficient as possible to write, transmit and store VFDs in order to remain profitable during the upcoming changes.

References

1. For a list of "medically-important" antibiotics from the FDA website, see Attachment 1. Last accessed on Oct 31, 2016 at http://www.fda.gov/AnimalVeterinary/DevelopmentApprovalProcess/ucm482107.htm.

2. See generally the Animal Drug Availability Act of 1996, which amended the Food Drug and Cosmetic Act of 1938 at 21 USC 9 §§ 301, 321, 354 and 360b.

3. It is important to note that a VFD is not analogous to a prescription and using the term "prescription" or "prescribe" raises additional requirements for the veterinarian found in many state veterinary practice acts and regulations. For example, Texas's Administrative Code allows disciplinary action for a veterinarian who "orders a prescription drug or controlled substance for the treatment of an animal without first establishing a veterinarian-client-patient relationship 4 Tex. Admin. Code § 801.402(13)

4. See 21 CFR § 558.6

5. See Attachment 2. Guidance for Industry Document # 209, last accessed Oct. 31, 2016 at http://www.fda.gov/downloads/animalveterinary/guidancecomplianceenforcement/guidanceforindustry/ucm216936.pdf. See also Attachment 3, Guidance for Industry Document #213, last accessed at Oct. 31, 2016 at http://www.fda.gov/downloads/animalveterinary/guidancecomplianceenforcement/guidanceforindustry/ucm299624.pdf.

6. *NRDC, Inc. v. FDA*, 760 F.3d 151 (2d Cir. 2014).

7. *Id.* at 154.

8. *Id.*

9. *Id.* at 156

10. See Guidance for Industry document #209 at 19.

11. Guidance for Industry Document #209.

12. Nancy E. Halpern, *Antibiotics in Food Animals: The Convergence of Animal and Public Health, Science,* Policy, Politics and the Law, 14 Drake J. Agric. L. 401 (2009); M.A. McCrackin, Kristi L. Helke, Ashley M. Galloway, Ann Z. Poole, Cassandra D. Salgado & Bernadette P. Marriott. *Effect of Antimicrobial Use in Agricultural Animals on Drug-resistant Foodborne Campylobacteriosis in Humans: A Systematic Literature Review*, Critical Reviews in Food Science and Nutrition, 56:13, 2115-2132 (2016).

13. Meghan Davis, Lainie Rutkow, *Regulatory Strategies To Combat Antimicrobial Resistance of Animal Origin: Recommendations for a Science-Based U.S. Approach*, 25 Tul. Envtl. L.J. 327 at 339. (2012).

14. *Id.* at 375. Discussion the ban of fluorquinolones in poultry and persistence of antibiotic resistance

15. Lisa Heinzerling. *The FDA's Continuing Incapacity on Livestock Antibiotics*, 33 Stan. Envtl. L.J. 325 (2014); Amanda Belanger, *A Holistic Solution for Antibiotic Resistance: Phasing out Factory Farms in Order to Protect Human Health*, 11 J. Health & Biomed. L. 145 (2015); Kayla Despenes, *A supersized solution to*

superbugs: Preventing a post-antibiotic era by optimizing consumer demand. 76 U. Pitt. L. Rev. 569 (2015); Nathalie Prescott, *Antibiotics: It's What's for Dinner*, 28 Geo. Envtl. L. Rev. 307 (2016); Lauren Orrico, *Squashing the superbugs: A proposed multifaceted approach to combatting antibiotic –resistant bacteria*, 27 J.L. & Health 259 (2014); Sarah Hagg. *FDA industry guidance targeting antibiotics used in livestock will not result in judicious use or reduction in antibiotic resistant bacteria*, 26 Fordham Envtl. Law Rev. 313 (2015); Nikki Sanford, *The battle against antimicrobial drug resistance: analyzing recent developments and the necessity for major agricultural reforms,* 40 Wm. & Mary Envtl. L. & Pol'y Rev. 989 (2016).

16. See Attachment 4 for summary sheets developed by stakeholders. Last accessed Oct 31, 2016. https://www.avma.org/KB/Resources/Pages/VFD123.aspx; http://www.fda.gov/AnimalVeterinary/DevelopmentApprovalProcess/ucm449019.htm http://www.farad.org/regulatory/vfd.asp; http://www.usfarad.org/vfd.html.

17. See Attachment 5 for a veterinary practice specific summary sheet.

18. 21 CFR § 558.6(b)(1).

19. See Attachment 6, which is continually updated at http://www.fda.gov/AnimalVeterinary/DevelopmentApprovalProcess/ucm460406.htm. Last accessed Oct 31, 2016.

20. The VCPR required federally states, "There is sufficient knowledge of the animal(s) by the veterinarian to initiate at least a general or preliminary diagnosis of the medical condition of the animal(s)." 21 § 530.3(i)(2)) The Federal rule goes on, "[s]uch a relationship can exist only when the veterinarian has recently seen and is personally acquainted with the keeping and care of the animal(s) by virtue of examination of the animal(s), and/or by medically appropriate and timely visits to the premises where the animal(s) are kept." 21 CFR § 530.3(i)(3).

21. *Id.*

22. Ultimately, the veterinarian's license to practice veterinary medicine is regulated under the state's practice act and subject to actions for issuing VFDs not incompliance with the state's VCPR requirements. See pg. 3 Interview with Michael Taylor FDA Deputy Commissioner for Foods and Veterinary Medicine and William Flynn Deputy Director for Science Policy in the FDA's Center for Veterinary Medicine at http://www.fda.gov/AnimalVeterinary/DevelopmentApprovalProcess/ucm448871.htm. Last Accessed Sep 15, 2016,

23. 21 CFR § 558.6 (b)(1) states the veterinarian must "be licensed to practice veterinary medicine."

24. See 21 CFR § 558.6(b)(3)(v) and Guidance for Industry Document #120 at 10.

25. Megan F. Davis, Lainie Rutkow, *Regulatory Strategies To Combat Antimicrobial Resistance of Animal Origin: Recommendations for a Science-Based U.S. Approach*, 25 Tul. Envtl. L.J. 327 at 339, 379. Meghan F. Davis et al., *An Ecological Perspective on U.S. Industrial Poultry Production: The Role of Anthropogenic Ecosystems on the Emergence of Drug-Resistant Bacteria from Agricultural Environments*, 14 Current Opinion Microbiology 244, 244-45 (2011).

26. 21 CFR § 558.3(b)(7).

27. Aureomycin® is a trademark of Zoetis Services, LLC which is not affiliated with the development of this guide, the reference is used illustration purposes only. https://www.zoetisus.com/products/beef/aureomycin_beef.aspx. Last Accessed Sept 15, 2106.

28. *Id.*

29. *Id.*

30. See the FDA's Fifth Annual Progress report at http://www.fda.gov/AnimalVeterinary/NewsEvents/CVMUpdates/ucm509403.htm. Last Accessed Sept 15, 2016.

31. 21 CFR § 558.6(b)(3)(xiii) repeats the prohibition on extralabel use of medicated feeds found at 21 CFR 530.11(b) and 21 USC § 360b(a) which states, "the following specific extralabel uses are not permitted... Extralabel use of an approved new animal drug or human drug in or on an animal feed."

32. 21 CFR § 558.6(a)(3)

33. Animals not kept for food producing purposes, such as pot-bellied pigs, and pet chickens are specially identified as "always considered to be food-producing animals" in other another FDA Guidance for Industry Document, #230 at 4. Last accessed on Oct, 31, 2016.

34. 21 CFR § 558.6(b)(3)(i-xv).

35. A 50 state survey of medical record retention, last accessed on Oct 31, 2016 at https://www.avma.org/Advocacy/StateAndLocal/Pages/sr-records-retention.aspx.

36. Principle of Veterinary Medical Records V.b. Available at https://www.avma.org/kb/policies/pages/principles-of-veterinary-medical-ethics-of-the-avma.aspx. Last accessed Sept. 15, 2016.

37. James F. Wilson, *Law and Ethics of the Veterinary Profession.* Priority Press, Yardley, PA. (1988).

38. 21 CFR § 530.3(i)

39. The federal VCPR states, "There is sufficient knowledge of the animal(s) by the veterinarian to initiate at least a *general or preliminary diagnosis* of the medical condition of the animal(s)." 21 CFR § 530.3(i)2.

40. *Id.*

41. The standard of care is framed as a question in most jurisdictions asking, "Did the veterinarian act as a ordinary and prudent practitioner in the situation?" A general practitioner will be held to the standard of care as other ordinary and prudent general practitioners and a specialist will be held to the standard of care of other similar specialist. For example a Diplomate of the American College of Swine veterinarians will be held to the standard of care of other Diplomates.

42. Many veterinary medical records for herd health currently identify the owner's herd as "cattle" or "swine" when the animals aren't uniquely identified. The medical record will often then follow medications dispensed chronologically and individually treated animals are identified when needed. This will pose problems when the different VFDs with different expiration dates are issued for certain lots of animals at the same time, and differing presumptive diagnoses are entered should the need arise to extract information about a specific VFD for specific animals.

43. The rule states specifically, "The veterinarian must send a copy of the VFD to the distributor via hardcopy, facsimile (fax), or electronically. If in hardcopy, the veterinarian must send the copy of the VFD to the

distributor either directly or through the client." 21 CFR § 588.6 (b)(ii)(8).

44. 21 CFR § 558.6(b)(3)

45. See Guidance for Industry Document #120 (D.4.) at page 13.

46. A true electronic VFD is required to be 21 CFR 11 compliant. This regulation lays out the verification and security requirements for a third party processor.

47. Several "stock" forms will be required for various different feeding scenarios such as

48. To be 21 CFR 11 compliant a veterinarian needs to make a one-time certification in paper form to the FDA and signed with a traditional handwritten signature. This letter needs to be mailed to the Office of Regional Operations (HFC-100), 5600 Fishers Lane, Rockville, MD 20857. This certification needs to state that the signatures are intended to be the legally binding equivalent of traditional handwritten signatures.

49. See FDA Guidance For Industry document #120 at 3.

50. See Guidance for Industry Document #230. Attachment 3.

51. See Attachment 7, which is replicated from the FDA example found in Guidance for Industry Document #233, pgs. 13,14, and 15.

52. See page 15 for electronic VFD requirements.

53. The National Agriculture Law Center will make further multimedia information, tailored for producers and distributers available.

54. Type A medicated articles are used to manufacture another Type A or for production of type B or C medicated feeds. Type B medicated feeds contain vitamins or minerals and are can be used for the manufacture of other medicated feeds and Type C medicated feeds are completed feeds or could be used as top dressing.

55. Guidance for Industry Document #120 at page 20.

56. 21 CFR § 558.3(b)(11) states that "An acknowledgement letter must be provided either in hardcopy or through electronic media, and must affirm: (1) that the distributor will not ship such VFD feed to an animal production facility that does not have a VFD; (2) that the distributor will not ship such VFD feed to another distributor without receiving a similar written acknowledgment letter; and (3) that the distributor has complied with the distributor notification requirements in 21 CFR 558.6(c)(5)."

57. The process to obtain a letter of distribution can be found at (21 CFR 558.6(c)(5)) and also within Guidance for Industry Document #120 at pg. 19. In summary, the distributor's complete name and business address, the distributor's signature or the signature of the distributor's authorized agent, and the date the notification was signed. The letter can be mailed or faxed to: Food and Drug Administration, Center for Veterinary Medicine, Division of Animal Feeds (HFV-220), 7519 Standish Pl., Rockville, MD 20855, FAX: 240-453-6882. Anytime the distributer moves, has a change in ownership or name, it must update the FDA within 30 days.

Section 11

ARAV Anesthesia

Bradley Waffa, MSPH, DVM;
James E. Bogan, Jr, DVM, DABVP (Canine and Feline Practice), CertAqV;
Marty Chu, DVM
Session Chairs

Pharmacokinetics of Meloxicam in Loggerhead (*Caretta caretta*), Kemp's Ridley (*Lepidochelys kempii*) and Green (*Chelonia mydas*) Sea Turtles after Subcutaneous Administration

Terry M. Norton, DVM, Dipl ACZM,
Tonya Clauss, DVM,
Sherry Cox, PhD

Session #2058

Affiliation: From Georgia Sea Turtle Center, Jekyll Island Authority, 214 Stable Road, Jekyll Island, GA 31527 USA (Norton); Georgia Aquarium , 225 Baker Street NW, Atlanta, GA 30313 (Clauss); College of Veterinary Medicine, University of Tennessee, Department of Biomedical and Diagnostic Science, Knoxville, TN 37996 USA (Cox)

Traumatic injuries (eg, boat strikes, shark bites), ingestion of fish hook and line, and entrapment in fishing gear are common in sea turtles that are presented for rehabilitation and often require surgery and pain management. The purpose of this study was to establish baseline pharmacokinetics of meloxicam (MLX) after a subcutaneous (SC) dose in 3 species of sea turtles: loggerhead, green and Kemp's ridley. One of the main differences among species is the rate of drug transformation. Due to the large degree of variation in metabolism, pharmacokinetic parameters must be established in each species. Plasma MLX concentrations considered to provide analgesia in mammals range from 0.5-1.5 μg/m1. At this time only preliminary data are available on 6 loggerhead, 7 Kemp's ridley, and 4 green sea turtles. At a dose of 2 mg/kg administered SC in loggerhead sea turtles, plasma MLX peaked at approximately 1 hour and maintained levels above 1 μ/ml for approximately 4 hours and then dropped below plasma MLX levels that would be considered analgesic. At a dose of 1 mg/kg MLX SC, Kemp's ridley sea turtles had peak plasma MLX levels at approximately 15 minutes and maintained plasma levels above 1 μg/ml for approximately 12 hours and had measurable levels for 48 hours. Most interestingly, green sea turtles given a dose of MLX of 1 mg/kg SC had peak concentrations at 30-60 minutes and maintained plasma levels above 1 μg/ml beyond 72 hours. Samples were only taken out to 72 hours, but will be extended as we move forward with the study.

Comparative Efficacy of Alfaxalone-Midazolam and Dexmedetomidine-Midazolam Sedation in Ball Pythons (*Python regius*)

Taylor J. Yaw, DVM,
Christoph Mans, Dr med vet, Dipl ACZM,
Grayson Doss, DVM,
Stephen Johnson, MS, MD, PhD,
Kurt Sladky, MS, DVM, Dipl ACZM, DipECZM (Herpetology)

Session #2070

Affiliation: From Department of Surgical Sciences and Comparative Biosciences, University of Wisconsin School of Veterinary Medicine, Madison, WI 53706, USA.

Ball pythons (*Python regius*) are one of the most commonly maintained pet reptile species. Sedation and anesthesia are frequently necessary to facilitate handling for physical examination and to enhance the quality and safety of diagnostic procedures. Systemically derived sedation efficacy data are lacking in many reptilian species, leaving clinicians to extrapolate information regarding drug dosages and safety from other, unrelated species. In a randomized, blinded complete crossover design, ball pythons (n = 12; 6 males and 6 females) were administered alfaxalone (5 mg/kg) + midazolam (0.5 mg/kg) and dexmedetomidine (0.05 mg/kg) + midazolam (0.5 mg/kg), subcutaneously in the cranial half of the body. Heart rate, respiratory rate, and sedation scores were monitored following drug administration. At 30 minutes post injection, endotracheal intubation and venipuncture were attempted, and at 45 minutes, the antagonists, flumazenil (0.05 mg/kg) and atipamezole (0.5 mg/kg), were administered subcutaneously in the cranial half of the body. Mean time to effect was 5.0 ± 0.0 minutes (alfaxalone-midazolam) and 7.5 ± 2.6 minutes (dexmedetomidine-midazolam), and recovery times were 38.3 ± 13.7 minutes (alfaxalone-midazolam) and 13.3 ± 3.3 minutes (dexmedetomidine-midazolam). Animals administered alfaxalone-midazolam lost muscle tone, while animals receiving dexmedetomidine-midazolam maintained moderate muscle tone. Animals in both groups were still reactive to endotracheal intubation. All animals recovered uneventfully from both sedation protocols. The results of this study provide additional sedation options for ball pythons.

Influence of Injection Site on Alfaxalone-Induced Sedation in Ball Pythons (*Python regius*)

Taylor J. Yaw, DVM,
Christoph Mans, Dr med vet, Dipl ACZM,
Grayson Doss, DVM,
Stephen Johnson, MS, MD, PhD,
Kurt Sladky, MS, DVM, Dipl ACZM, DipECZM (Herpetology)

Session #2117

Affiliation: From the Department of Surgical Sciences and Comparative Biosciences, University of Wisconsin School of Veterinary Medicine, Madison, WI 53706, USA.

Sedation and anesthesia are frequently necessary to facilitate handling for physical examination, and to enhance the quality and safety of diagnostic procedures in snake species, such as ball pythons (*Python regius*). Further, the hepatic-first pass effect and reduced drug efficacy, following injection of drugs in the caudal body half, is of concern in reptiles. It has been demonstrated in turtles, geckos and crocodilians that injection in the hind limbs results in significantly reduced efficacy of anesthetic drugs, which undergo hepatic metabolism; however, this phenomenon has not been investigated in any snake species. In a randomized, blinded complete cross-over design, ball pythons (n = 8; 4 males and 4 females) were administered alfaxalone (5 mg/kg) subcutaneously in the cranial and caudal half of the body. Heart rate, respiratory rate, and sedation scores were monitored following drug administration. Alfaxalone administered in the cranial half of the body resulted in significantly ($P < 0.05$) lower heart rate, respiratory rate, and significantly higher sedation scores, compared to administration in the caudal body half. The results of this study demonstrate the importance of injection site for drug administration in snakes, and provide a safe sedation option for ball pythons. This study confirms that drugs, which undergo hepatic metabolism, should not be administered in the caudal body half of snakes.

Dexmedetomidine-Midazolam versus Alfaxalone-Midazolam Sedation in Leopard Geckos (*Eublepharis macularius*)

Christoph Mans, Dr med vet, Dipl ACZM,
Grayson Doss, DVM,
Dustin Fink, BS,
Kurt K. Sladky, MS, DVM, Dipl ACZM, DipECZM (Herpetology)

Session #2123

Affiliation: From the School of Veterinary Medicine, University of Wisconsin-Madison, 2015 Linden Drive, Madison, WI 53706, USA.

The goal of this study was to develop safe, clinically applicable sedation protocols in leopard geckos using dexmedetomidine-midazolam and alfaxalone-midazolam combinations. Nine healthy, adult leopard geckos were administered dexmedetomidine (0.1 mg/kg) and midazolam (1.0 mg/kg) (D-M group) or alfaxalone (15 mg/kg) and midazolam (1.0 mg/kg) (A-M group) subcutaneously in a randomized, blinded, complete crossover study. Physiologic parameters and multiple reflexes were assessed. Reversals were administered 45 minutes after initial injection. Heart rate but not respiratory rate decreased significantly over time in both groups. Heart rate was significantly higher for geckos in the A-M group for all time points. Sedation scores between the two protocols were similar. Mean ± standard deviation (SD) recovery for the D-M group was 6.1 ± 2.2 min; recovery was significantly longer for the A-M group at 56 ± 29 min. This study describes two safe and effective anaesthetic protocols for use in leopard geckos. Results demonstrate that combinations of dexmedetomidine-midazolam and alfaxalone-midazolam provide adequate immobilization for performing physical examinations, minor procedures like venipuncture, and diagnostic imaging in leopard geckos. Recovery is rapid following reversal of dexmedetomidine-midazolam anaesthesia in this species and alfaxalone-midazolam provides a more rapid onset of anaesthesia, but with significantly longer recovery times.

Effect of Injection Site on Dexmedetomidine-Ketamine Induced Sedation in Leopard Geckos (*Eublepharis macularius*)

Christoph Mans, Dr med vet, Dipl ACZM,
Dustin Fink, BS,
Grayson Doss, DVM,
Kurt K. Sladky, MS, DVM, Dipl ACZM, DipECZM (Herpetology)

Session #2124

Affiliation: From the School of Veterinary Medicine, University of Wisconsin-Madison, 2015 Linden Drive, Madison, WI 53706, USA.

Location for parenteral drug administration is a controversial topic in reptile medicine.[1] Previous investigations in turtles[2-4] demonstrated that administration of compounds in the caudal body half that are hepatically transformed or metabolized can result in reduced drug levels or lack of efficacy, secondary to a hepatic first pass effect. The goal of this study was to investigate the effect of injection site location on dexmedetomidine-ketamine anaesthesia in leopard geckos. Nine healthy, adult leopard geckos were administered dexmedetomidine (0.1 mg/kg) and ketamine (10 mg/kg) intramuscularly in either the forelimb (FL) or hind limb (HL). Physiologic parameters and reflexes were assessed. Righting reflex was absent in 7/9 animals following FL injection and none of the animals following HL injection. Onset of sedation was observed as soon as 5 min in the FL group; onset was not observed in the HL group for at least 10 min. Median and maximum sedation scores were higher and heart rate significantly lower for the FL group. Intramuscular dexmedetomidine-ketamine in the HL of leopard geckos is significantly less effective than FL injection. Since both drugs undergo hepatic metabolism and the hind limbs drain directly into the liver via the ventral abdominal vein, a hepatic first pass effect is the likely cause for the reduced efficacy. Caution should be used when administering compounds that undergo hepatic metabolism or transformation in the caudal body half in reptiles.

References

1. Holz P, Barker IK, Burger JP, et al. The effect of the renal portal system on pharmacokinetic parameters in the red-eared slider (*Trachemys scripta elegans*). *J Zoo Wildl Med.* 1997;28:386-393.

2. Scheelings TF. Use of intravenous and intramuscular alfaxalone in Macquarie river turtles (*Emydura macquarii*). *J Herpetol Med Surg.* 2013;23:91-94.

3. Lahner LL, Mans C, Sladky KK. Comparison of anesthetic induction and recovery times after intramuscular, subcutaneous or intranasal dexmedetomidine-ketamine administration in red-eared slider turtles (*Trachemys scripta elegans*). *Proc Annu Conf Am Assoc Zoo Vet.* 2011;136-137.

4. Kummrow MS, Tseng F, Hesse L, Court, M. Pharmacokinetics of buprenorphine after single-dose subcutaneous administration in red-eared sliders (*Trachemys scripta elegans*). *J Zoo Wildl Med.* 2008;39:590-595.

Section 12

ARAV Chelonians

Shane Simpson, BVSc(Hons), GCM(VP), CMAVA and
Louisa M. Asseo, DVM, DABVP (Canine and Feline Practice)
Session Chairs

Comparison of Capillary Zone Electrophoresis in Two Tortoise Species, *Testudo hermanni* and *Testudo horsfieldii*

Christoph Leineweber, med vet,
Anke C. Stöhr, med vet, ZB Reptilien,
Sabine Öfner, med vet FTÄ Reptilien,
Karina Mathes, Dr med vet, DipECZM (Herpetology), FTÄ Reptilien,
Rachel E. Marschang, PD Dr med vet, DipECZM (Herpetology), FTÄ Mikrobiologie, ZB Reptilien

Session #2035

Affiliation: From Laboklin GmbH & Co. KG, Steubenstr. 4, Bad Kissingen, Germany (Leineweber, Marschang), Kleintierklinik Egelsbach, Boschring 8, Egelsbach, Germany (Stöhr), Reptile Rescue Center Munich e.V., Kaulbachstraße 37, Munich, Germany (Öfner), Clinic for Small Mammals, Reptiles and Birds, University of Veterinary Medicine Hannover, Foundation, Hannover, Germany (Mathes, Leineweber).

Capillary zone electrophoresis for detection of protein imbalances is faster, less expensive and the separation of the protein fractions is more detailed than other methods. At this time no reference values for protein electrophoresis measured by capillary zone electrophoresis are available for chelonians or other reptiles. Blood samples were obtained from 10 adult healthy Hermann's tortoises (*Testudo hermanni*) and 10 adult healthy Russian tortoises (*Testudo horsfieldii*) in June 2016. The blood was collected from the dorsal coccygeal or subcarapacial vein and filled into lithium-heparin tubes. The centrifuged samples were sent to the veterinary laboratory and were analyzed 1 day after collection. Plasma total protein was determined using the cobas® 6000 analyzer series (Roche Diagnostics, Germany). Plasma capillary zone electrophoresis was performed using the sebia MINICAP® (Sebia, France). The results show that the protein profiles are very different between the individual chelonian species. Russian tortoises have a triple albumin peak, a lower alpha-globulin peak and in the mean a higher albumin:globulin ratio as the Hermann´s tortoises in June. The establishment of standard ranges during different seasons and species will facilitate the diagnosis of diseases in chelonians. The few publications available on electrophoresis by cellulose acetate electrophoresis (CAE) and agarose gel electrophoresis (AGE) in reptiles indicate that this is a good tool for the detection of protein imbalances caused by certain pathologic conditions (eg, infectious diseases such as parasites or viral infections). Sampling over the course of 1 year and of other species also is planned.

Intranuclear Coccidia of Tortoises in Europe

Ekaterina Kolesnik, med vet,
Janosch Dietz, med vet,
Kim O. Heckers, Dr med vet, FTA Pathologie,
Rachel E. Marschang, PD Dr med vet, DipECZM (Herpetology), FTÄ Mikrobiologie, ZB Reptilien

Session #2036

Affiliation: From Laboklin GmbH & Co. KG, Steubenstr. 4, 97688 Bad Kissingen, Germany.

Intranuclear coccidia are important pathogens in a wide range of chelonian species, where they have been associated with systemic disease. Multiple reports have described these pathogens in tortoises and a few turtle species in North America. Reports from other parts of the world are rarer. Using a real-time PCR, samples from 1011 chelonians were screened for the presence of intranuclear coccidia of tortoises (TINC). Samples included mostly oral swabs and nasal washes as well as tissue samples in a few cases. Most samples were obtained from captive chelonians in various European countries. Coccidia were detected in a total of 27 chelonians (2.7%), including the first description of TINC in marginated tortoises (*Testudo marginata*), Hermann's tortoises (*T. hermanni*), African spurred tortoises (*Centrochelys sulcata*) and a yellow-footed tortoise (*Chelonoidis denticulatus*). TINC also were detected in radiated tortoises (*Astrochelys radiata*), leopard tortoises (*Stigmochelys pardalis*), Indian star tortoises (*Geochelone elegans*), and a bowsprit tortoise (*Chersina angulata*). Radiated tortoises, leopard tortoises, and Indian star tortoises were each significantly more often positive than Hermann's tortoises. Positive samples were obtained from chelonians in Spain, Germany, Switzerland, Denmark, the United Kingdom, Sweden, Italy, France, Portugal, Austria, and China.

Disclosure Statement: The authors are employed by LABOKLIN, Bad Kissingen, Germany.

Serial Assessment of Plasma Corticosterone and Free Thyroxine Concentrations during Rehabilitation of 85 Cold-stunned Kemp's Ridley Turtles (*Lepidochelys kempii*), with Preliminary Observations on Exogenous Thyroxine Supplementation

Charles Innis, VMD, DABVP (Reptile and Amphibian Practice),
Kathleen Hunt, PhD,
Constance Merigo, BS,
Rosalind M. Rolland, DVM

Session #2040

Affiliation: From New England Aquarium, Central Wharf, Boston, MA 02110, USA.

Plasma corticosterone and free thyroxine (fT4) were serially monitored using validated assays for 85 juvenile Kemp's ridley turtles (*Lepidochelys kempii*) during hospitalization after cold-stunning. Corticosterone concentrations upon admission were extremely high (mean 58 ng/ml), but dropped steeply by 2 weeks of rehabilitation and typically remained < 5 ng/ml thereafter. Elevated corticosterone concentrations later in rehabilitation correlated with reduced growth rate and increased mortality. Free thyroxine was extremely low at admission, but gradually increased to 3-4 pg/ml after 3 to 5 months of rehabilitation. Free thyroxine concentration at 2 weeks of rehabilitation correlated with food intake. Levothyroxine supplementation (0.03 mg/kg PO q48h; Soloxine, Virbac AH Inc, Fort Worth, TX, USA) for 4 anorexic turtles raised fT4 levels; however, most plasma samples obtained from turtles during supplementation had fT4 concentrations that were still low compared to eventual convalescent fT4 concentrations. Four initially anorexic turtles given supplemental thyroxine had increased growth rates compared to 5 anorexic unsupplemented turtles even after discontinuation of supplementation, but there was no difference in time to first food acceptance. It is likely that severely increased corticosterone concentrations may contribute to the high incidence of infection in cold-stunned Kemp's ridley turtles. Thus, clinicians should take precautions to minimize turtle stress during transportation, examination, and handling. Detection of abnormal corticosterone and thyroxine concentrations may attract clinical attention to a patient for which other diagnostic modalities had not detected a problem. Further refinement of thyroxine supplementation recommendations for this species will require additional pharmacokinetic and clinical studies.

Acknowledgments: This study was funded by the Morris Animal Foundation (D13ZO-022), and approved by the New England Aquarium Animal Care and Use Committee (2012-04). We thank the staff and volunteers of the Massachusetts Audubon Society Wellfleet Bay Sanctuary, and the New England Aquarium Animal Health and Rescue and Rehabilitation Departments for recovery, clinical care, and sample preparation for these turtles. Adam Kennedy was particularly important in ensuring timely sample collection. Sea turtle rehabilitation at New England Aquarium is conducted with authorization of the United States Department of Commerce, National Oceanic and Atmospheric Administration, National Marine Fisheries Service and the United States Department of the Interior Fish and Wildlife Service (permit number TE-697823).

Evaluation of Health in Texas Tortoises (*Gopherus berlandieri*) Via Multiple Patient Side Analytes

Michelle Taylor, and
J. Jill Heatley, DVM, MS, DABVP (Avian Practice), DABVP (Reptile and Amphibian Practice), Dipl ACZM

Session #2041

Affiliation: From Texas A&M University, College of Veterinary Medicine and Biomedical Sciences, University Dr, College Station, TX 77843, USA.

The Texas tortoise, *Gopherus berlandieri*, is the only native tortoise species in Texas and is considered threatened. Little is known regarding reptilian inflammatory response and markers of overall health. References conflict regarding the viability of many markers of inflammation in reptiles and few studies have assessed these parameters in chelonians proven to be suffering inflammation.[1-5] *Mycoplasma* spp. and tortoise herpesvirus 2 (TeHV2) exposed Texas Tortoises (n = 10) and apparently healthy male Texas Tortoises (n = 10) maintained at the Texas A&M University, College of Veterinary Medicine and Biomedical Sciences, Winnie Carter Exotic Animal Center were sampled via jugular venipuncture to determine multiple analytes for assessment of tortoise health. Whole blood was used to determine erythrocyte sedimentation rate ESR, fibrinogen, lactate, packed cell volume and total solids via patient side methodologies while lactic acid, total protein, albumin, globulin (indirectly), uric acid and iron were determined from plasma using a standard laboratory analyzer (Vitros ® 4600, Ortho-clinical Diagnostics, Raritan NJ, USA). While plasma fibrinogen was significantly different based on health status the unexpected lesser values found during determination of fibrinogen in this species suggest that in contrast to birds and mammals, Texas tortoise refractometric index of plasma may decrease in response to heating and centrifugation. In *Mycoplasma* and TeHV exposed tortoises, ESR was higher, and plasma fibrinogen and lactic acid were lower than in apparently healthy tortoises. Fair agreement of lactate and lactic acid, as well as total solids and total protein suggests that patient side methods (lactometer and refractometer) are acceptable for clinical use in these species.

References

1. Moore AR, Allender MC, Mitchell MA, MacNeill AL. Evaluation of plasma fibrinogen concentration as a diagnostic indicator of inflammation in red-eared sliders (*Trachemys scripta elegans*). *J Am Vet Med Assoc.* 2015;246(2):245-253.

2. Moore AR, Allender MC, MacNeill AL. Effects of ranavirus infection of red-eared sliders (*Trachemys scripta elegans*) on plasma proteins. *J Zoo Wildl Med.* 2014;45(2):298-305.

3. Naldo JL, Libanan NL, Samour JH. Health assessment of a spiny-tailed lizard (*Uromastyx* spp.) population in Abu Dhabi, United Arab Emirates. *J Zoo Wildl Med.* 2009;40(3):445-452.

4. Parkinson L, Olea-Popelka F, Klaphake E, Dadone L, Johnston M. Establishment of a fibrinogen reference interval in ornate box turtles (*Terrapene ornata ornata*). *J Zoo Wildl Med.* 2016;47(3):754-759.

5. Zhou X, Wang L, Feng H, et al. Acute phase response in Chinese soft-shelled turtle (*Trionyx sinensis*) with *Aeromonas hydrophila* infection. *Dev Comp Immunol.* 2011;35(4):441-451.

The Microbiome of Sea Turtles

T. Franciscus Scheelings, BVSc, MVSc, MANZCVSc (Med Aust Wildl), DipECZM (Herpetology), Marcel Klaassen, PhD, Richard Reina, PhD

Session #2139

Affiliation: From School of Biological Sciences, Monash University, Wellington Rd, Clayton, VIC, Australia, 3800 (Scheelings and Reina) and Centre for Integrative Ecology, Deakin University, Waurn Ponds, VIC, Australia, 3216 (Klaassen).

Residing on and within every metazoan species is a diverse array of microbial organisms collectively known as the microbiome. These microbial inhabitants have played a vital role in driving evolution and shaping ecology. The microbiome has an intimate relationship with its host, and is fundamental in a range of normal physiological processes. In fact, without this symbiotic relationship, life as we know it would not be possible. Little is known about the microbiome in reptiles and how it infers fitness to its hosts. The microbiome acquisition plays an important role in sea turtle conservation.

Gap in pagination due to unavailable material.

Pages 154-194

Section 14

ARAV Master Classes

Anneliese Strunk, DVM, DABVP (Avian Practice);
Mike Miller, DVM;
Karina Mathes, Dr med vet, Cert Spec in Reptiles and Amphibians, DipECZM (Herpetology)
Session Chairs

Reptile Sedation, Anesthesia, and Analgesia

Christoph Mans, Dr med vet, Dipl ACZM

Session #2128

Affiliation: From the School of Veterinary Medicine, University of Wisconsin-Madison, 2015 Linden Drive, Madison, WI 53706, USA.

Introduction

Utilizing chemical restraint methods in reptile patients is commonly needed for diagnostic and therapeutic procedures; however, the significant differences in anatomy and physiology amongst different groups of reptiles as well as several other factors such as body temperature, injection site and route as well as underlying disease processes make sedation and anesthesia in reptiles challenging at times.[1,2]

Reducing environmental temperatures below the preferred temperature range of reptile patients will lead to a delay in onset of action of anesthetic drugs, longer action and delayed recovery. If temperatures are increased, onset of sedation or anesthesia is more rapid, but the duration of effect can be shorter. Therefore it is critical to ensure that reptile patients are maintained at an appropriate temperature prior to and during sedation and anesthesia.

Injection of sedative and anesthetic drugs in the caudal body-half should be avoided if possible, since some drugs may undergo a hepatic first pass effect, which may be more important for most anesthetic drugs, compared to the more often discussed renal first-pass effect.[1,3] Plasma levels and efficacy of anesthetic drugs can be greatly reduced and may require repeated drug administration.[3]

Historically the intramuscular route has been recommended as the route of choice for non-vascular administration of anesthetic and sedative drugs in reptiles; however, the subcutaneous route provides a suitable alternative, in particular in animals with reduced muscle mass or if larger volumes of drugs need to be administered. While the onset of effect is often slower following subcutaneous administration, the depth of sedation or anesthesia reached is ultimately similar following intramuscular administration.[1]

By considering short acting drugs and partial or completely reversible drug protocols for sedation (Table 1) and anesthesia (Table 2), prolonged recoveries or fatalities can be avoided. Sedation is often sufficient for most diagnostic procedures as well as certain therapeutic procedures, particularly if combined with local anesthetics.[2] Lidocaine and bupivacaine can be used for local infiltration nerve block as well as for spinal anesthesia in reptiles.[2,4]

Table 1. Sedation protocols commonly used in reptiles. Atipamezole is used to antagonize medetomidine/dexmedetomidine in 1:1 volume, SC or IM. Flumazenil is used to antagonize midazolam: 0.05 mg/kg SC, IM or IV. Nalaxone is used to antagonize morphine or hydromorphone: 0.04 mg/kg SC, IM, or IV.[a]

Protocol	Dose (mg/kg)	Route	Species	Comments
Chelonians				
Midazolam	2	SC/IM	Red-eared slider	Mild sedation, inconsistent effects, completely reversible
Dexmedetomidine + midazolam	0.1 + 1	SC/IM	Red-eared slider NA box turtles	Mild to moderate sedation, completely reversible; rapid recovery after reversal
Dexmedetomidine + midazolam + ketamine	0.1 + 1 + 2	SC/IM	Red-eared slider	Moderate sedation, partially reversible; rapid recovery after reversal
Medetomidine + ketamine	0.1 + 5	SC/IM	Red-eared slider	Deep sedation, suitable for intubation
Medetomidine + ketamine + morphine	0.15 + 2.5 + 1	SC/IM	Tortoises	Deep sedation - light anesthesia, partially reversible; rapid recovery after reversal
Medetomidine + midazolam + morphine	0.2 + 2 + 1	SC/IM	Tortoises	Deep sedation - light anesthesia, completely reversible; rapid recovery after reversal.
Alfaxalone	10	IM	Red-eared slider	Moderate sedation, suitable for intubation
Propofol	2-5	IV		Moderate sedation - light anesthesia
Lizards				
Dexmedetomidine + midazolam	0.1 + 1	SC/IM	Bearded dragon, green iguana	Moderate sedation, completely reversible; rapid recovery after reversal
Dexmedetomidine + midazolam + ketamine	0.05 - 0.1 + 1 + 3	SC-IM	Bearded dragon, green iguana	Moderate - deep sedation, partially reversible; rapid recovery after reversal
Alfaxalone	10	IM, IV	Green iguana	Light-moderate sedation, maximum effect after ~ 5 min, duration of effect ~ 10 min
Propofol	3-5	IV, IO		Deep sedation - light anesthesia
Snakes				
Midazolam	1-2	IM		Mild sedation, inconsistent effects
Telazol	2-5	SC, IM	Large snakes	Mild-moderate sedation, endotracheal intubation
Ketamine	5-10	SC, IM		Mild-moderate sedation, endotracheal intubation
Propofol	3-5	IV		Moderate sedation - light anesthesia

[a]Abbreviations: SC=subcutaneous; IM=intramuscular; IV=intravenous; IO=intraosseous.

Table 2. Anesthetic protocols commonly used in reptiles. Atipamezole is used to antagonize medetomidine/dexmedetomidine in 1:1 volume, SC or IM. Flumazenil is used to antagonize midazolam: 0.05 mg/kg SC, IM or IV. Nalaxone is used to antagonize morphine or hydromorphone: 0.04 mg/kg SC, IM, or IV.[a]

Protocol	Dose (mg/kg)	Route	Species	Comments
Chelonians				
Medetomidine + midazolam + ketamine + hydromorphone	0.15 + 0.5 + 10 + 0.5	SC/IM	Tortoises	Surgical anesthesia, maintain with gaseous anesthetic if necessary
Medetomidine + ketamine	0.2 + 10	SC/IM	Red-eared slider	Surgical anesthesia partially reversible
Medetomidine + ketamine + morphine	0.1 + 10 +1.5	SC/M	Map turtle	Surgical anesthesia, Administration of nalaxone recommended if recovery is slow or to prevent renarcotization
Dexmedetomidine + ketamine + hydromorphone	0.15 + 10 + 0.5	SC/IM		Surgical anesthesia, supplement with gaseous anesthesia if necessary
Propofol	2-10	IV		Induction agent; use lower dose in large tortoises
Propofol	10-20	IV	Red-eared slider	Light anesthesia for ~ 60 min (10mg/kg) or ~ 90 min (20mg/kg)
Alfaxalone	20	IM	Red-eared slider	Induction agent; light anesthesia for 28 ± 13 min at 20°C
Lizards				
Alfaxalone	20-30	IM	Green iguana	Induction agent, effect after 5 minutes, duration of effect 20-40 min
Alfaxalone	10	IV	Green iguana	Induction agent; maintain with gaseous anesthetic
Propofol	5-10	IV, IO	Green iguana	Induction agent, apnea common, requires intubation and IPPV
Snakes				
Propofol	3-10	IV, IC		Induction agent; maintain with gaseous anesthetic
Alfaxalone	9	IV	Australian snakes	Light anesthesia, endotracheal intubation
Telazol®	2-6	IM		Induction agent; maintain with gaseous anesthetic
All reptiles				
Isoflurane	2-5%			Induction 5%, maintenance 2-3%
Sevoflurane	2.5-8%			Induction 7-8%, maintenance 2.5-4.5%

[a]Abbreviations: SC=subcutaneous; IM=intramuscular; IV=intravenous; IO=intraosseous.

Spinal Anesthesia and Analgesia

Intrathecal (subdural) administration of anesthetic and analgesic drugs in turtles and tortoises is a novel technique for the induction of spinal anesthesia and analgesia. Possible indications for spinal anesthesia include surgical procedures of the tail, phallus, cloaca, and hindlimbs. Intrathecal injections are performed at the level of the coccygeal vertebrae. The intrathecal administration of lidocaine (2% preservative-free, 4 mg/kg) and bupivacaine (0.5% preservative free, 1 mg/kg) provides regional anesthesia of the tail, cloaca and hindlimbs for about 1 and 2 hours respectively. The intrathecal administration of morphine (preservative free, 0.1- 0.2 mg/kg) with lidocaine provides regional analgesia for up to 48 hours. Strict aseptic techniques should be used to avoid iatrogenic complications, and only preservative-free drugs should be injected into the intrathecal space so that spinal toxicity and secondary neurologic complications are avoided.

Reptile Analgesia

Our knowledge on reptile analgesia continues to grow, and increasing scientific evidence is available to guide the selection of drugs, which are likely to provide analgesia in reptiles.[5,6] The side effects of opioid analgesics, such as sedation and respiratory depression and well as differences in response among groups of reptiles should be carefully considered. In general most mu-opioid receptor agonists have been shown to provide analgesia in turtles and tortoises as well as lizards and crocodilians. Drugs such as morphine, hydromorphone, fentanyl and tramadol are most commonly used.[5,7,8] Limited information is available in regards to drug side effects across different reptile species. Drugs previously believed to have analgesic properties in reptiles, such as butorphanol and buprenorphine, have so far not shown to be effective in providing analgesia in reptiles (Table 3).[8,9] Non-steroidal anti-inflammatory drugs are frequently administered to reptiles; however, to date, no evidence for analgesic efficacy of these drugs in reptiles exists. Meloxicam appears to be safe and even at high doses no adverse effects were found.[10] Future research is necessary in order to determine safe and effective analgesic drug protocols for the reptile species most commonly kept in captivity.

Table 3. Analgesic drugs commonly used in reptiles.

Drug	Dose (mg/kg)	Route	Species	Comments
Hydromorphone	0.5-1	SC/IM	Chelonians	q24h
Morphine	1-2	SC/IM	Chelonians	q24h
Tramadol	5-10	PO, SC/IM	Chelonians	q48-72h
Buprenorphine	0.02-0.2	SC/IM	Chelonians	No analgesia demonstrated
Butorphanol	1-8	SC/IM	Chelonians	No analgesia demonstrated
Meloxicam	0.2	PO, SC/IM	Green iguanas	q24h, pk data only
Ketoprofen	2	SC/IM	Green iguanas	> q24h, pk data only

References

1. Sladky KK, Mans C. Clinical anesthesia in reptiles. *J Exot Pet Med.* 2012;21:17-31.

2. Schumacher J, Mans C. Anesthesia. In: Mader DR, Divers SJ, eds. *Current Therapy in Reptile Medicine and Surgery*. St Louis, MO: Elsevier;2014.

3. Kummrow MS, Tseng F, Hesse L, et al. Pharmacokinetics of buprenorphine after single-dose subcutaneous administration in red-eared sliders (*Trachemys scripta elegans*). *J Zoo Wildl Med.* 2008;39:590-595.

4. Mans C. Clinical technique: intrathecal drug administration in turtles and tortoises. *J Exot Pet Med.* 2014;23:67-70.

5. Sladky KK. Analgesia In: Mader DR, Divers SJ, eds. *Current Therapy in Reptile Medicine and Surgery.* St Louis, MO: Elsevier;2014.

6. Sladky KK, Mans C. Clinical analgesia in reptiles. *J Exot Pet Med.* 2012;21:158-167.

7. Giorgi M, Salvadori M, De Vito V, et al. Pharmacokinetic/pharmacodynamic assessments of 10 mg/kg tramadol intramuscular injection in yellow-bellied slider turtles *(Trachemys scripta scripta)*. *J Vet Pharmacol and Ther.* 2015. In press.

8. Mans C, Lahner LL, Baker BB, et al. Antinociceptive efficacy of buprenorphine and hydromorphone in red-eared slider turtles *(Trachemys scripta elegans)*. *J Zoo Wildl Med.* 2012;43:662-665.

9. Sladky KK, Miletic V, Paul-Murphy J, et al. Analgesic efficacy and respiratory effects of butorphanol and morphine in turtles. *J Am Vet Med Assoc.* 2007;230:1356-1362.

10. Divers SJ, Papich M, McBride M, et al. Pharmacokinetics of meloxicam following intravenous and oral administration in green iguanas *(Iguana iguana)*. *Am J Vet Res.* 2010;71:1277-1283.

Antiviral Therapy and Prophylaxis in Reptile Medicine

Rachel E. Marschang, PD Dr vet med, DipECZM (Herpetology), FTÄ Mikrobiologie, ZB Reptilien

Session #2037

Affiliation: From Laboklin GmbH & Co. KG, Steubenstr. 4, 97688 Bad Kissingen, Germany.

Abstract: Dealing with viral infections in reptiles, as in other animals, can be frustrating. In reptiles especially, there is a lack of specific antiviral agents that have been shown to be effective against specific viruses of reptiles. Pharmocokinetic studies also are lacking. There are several strategies that can be used to deal with viral infections in reptiles. These include treatment with antiviral agents, modulation of the environment, especially the temperature, to inhibit virus replication and facilitate host immune response, immune modulation, vaccination, treatment of secondary infection, and disinfection. It is important to note that viral infections often persist in reptiles and treatment of clinical illness will not generally result in freedom from infection.

Introduction

This presentation aims to give an overview of methods that have been described or suggested to help treat clinical viral disease in reptiles. They include the use of antiviral agents which inhibit virus replication in the host, as well as the use of temperature to inhibit virus replication and increase host immune response. The local application of disinfectants also has been described to treat lesions caused by viruses (eg, in the case of viral stomatitis or by inhalation in viral upper respiratory tract disease). Agents that modulate the immune system have also been suggested for possible use in viral diseases in reptiles, while in some cases, vaccination may help reduce the effects of a viral infection or lead to faster recovery. Vaccination is also a possible prophylactic measure against viral infections, but has only rarely shown promise for use in reptile medicine. Secondary infections may play a major role in the development and course of disease during viral infections, making optimization of husbandry and treatment of opportunistic pathogens an important aspect of clinical disease control. Finally, disinfection plays a major part in infection control in medicine in general, and understanding best practices and what can affect successful disinfection is important in developing a hygiene plan and preventing spread of infection both in the animal's home and in veterinary practice.

Antiviral Agents

The rate at which antiviral agents have been discovered has rapidly increased over the past years. In reptile medicine, however, there has been very little work done on the efficacy, safety, or pharmacokinetics of antiviral substances. The most commonly described antiviral in reptiles is acyclovir, which has been used to treat animals with herpesvirus infections. Acyclovir is a nucleoside analogue which inhibits the DNA polymerase. The most commonly described dose is 80 mg/kg PO in chelonians,[1,2] although there is some indication that this does not elicit sufficient serum concentrations and that the half-life may be short. Topical acyclovir use to treat herpesviral oral lesions in tortoises has also been described.[3] Related nucleoside analogues, such as valacyclovir, cidofovir, or famciclovir may be more effective and safer, although little information is available on their use in reptiles. These substances have also been used to treat infections with other DNA viruses than herpesviruses in humans and animals, although acyclovir and cidofovir were not effective against a ranavirus in vitro.[4] Other substances anecdotally used to treat viral infections in reptiles include lysine and ribavirin.

Temperature Regulation

The viruses that infect reptiles are generally able to replicate within a defined temperature range, and environmental temperature can affect the rate of replication. In addition, the immune system of reptiles is affected by temperature, and environmental temperature will affect the animal's ability to react to a viral infection as well as to secondary infectious agents. It has therefore been suggested that adjustment of environmental temperature can be used as part of a treatment regimen for viral infections in reptiles (eg, to treat ranavirus infections) since ranaviruses are unable to replicate at temperatures higher than 32 °C. It is important to keep the preferred optimal temperature zone of the species in question in mind when developing a treatment plan that includes adjusting the environmental temperature so as not to harm the animal.

Immune Modulation

In addition to temperature, chemical substances can be used to influence the immune system and to try to provoke an increased reaction to infection. In reptile medicine, the use of inactivated parapox ovis virus (Zylexis, Pfizer AG, Zurich, Switzerland) has been suggested for this purpose, although the efficacy has not been proven.

Treating Secondary Infections

Bacteria, fungi, parasites, and other viruses can strongly influence the clinical course of a viral infection and vice versa. Diagnosing and treating multiple infections can therefore be an important part of a therapy plan. In addition, animals may need general supportive therapy and nursing during acute disease.

Vaccination

The use of vaccines to protect reptiles from viral infection is not well developed. Studies attempting to vaccinate tortoises against inactivated herpesviruses and snakes against inactivated ferlaviruses have not been promising, although they have not been associated with adverse effects, either.[5,6] In crocodilians, on the other hand, the use of vaccination has shown good results, with the use of an autogenous poxvirus vaccine leading to reduced scarring.[7] A commercial vaccine against West Nile virus has been provisionally licensed for use in alligators in the United States (Boehringer Ingelheim, St. Joseph, MO).

Disinfection

General hygiene and specifically disinfection play an important role in the prevention of infection. Chemical and physical methods are routinely used for the inactivation of viruses in animal husbandry and in veterinary practice. In addition, biological inactivation of viruses can play a role in infection dynamics in some cases (eg, for animals kept in outdoor enclosures). As a general rule, enveloped viruses are more easily inactivated than non-enveloped viruses. In some cases, double stranded DNA viruses may be more stable against some inactivation methods than other viruses. Viruses found in reptiles are listed in Table 1 according to some of their physical properties. In reptile medicine, disinfectants can be used to lower the number of infectious virions in the environment as well as to locally treat lesions caused by viral infections. In the latter case, treatment can aim to inactivate the viruses involved as well as to inactivate secondary infectious agents (eg, bacteria and fungi). Disinfectants reportedly used to treat viral lesions include the use of chlorhexidine (0.5% solution) to treat herpesviral stomatitis and glossitis in tortoises and use of the disinfectant F10 to treat upper respiratory tract disease.[8]

Table 1. Virus families and genera described in orders of reptiles according to structural properties relevant for antiviral agents and virus disinfection.

Virus family	Virus genus	Non-avian reptile host orders in which viruses in this family have been described
Enveloped viruses		
dsDNA		
Hepadnaviridae	Unassigned	Testudines, Squamata
Herpesviridae	*Scutavirus*	Testudines
	Unassigned	Squamata, Crocodylia
Iridoviridae[a]	*Ranavirus*	Testudines, Squamata
	Iridovirus	Squamata
	"Hemocytivirus	Testudines, Squamata
Poxviridae	*Crocodylipoxvirus*	Crocodylia
	Unclassified	Testudines, Squamata
ssRNA		
Arenaviridae	*Reptarenavirus*	Squamata
Bornaviridae	Unassigned	Squamata
Bunyaviridae	*Orthobunyavirus*	Testudines
	Unassigned	Squamata
Coronaviridae	Unassigned	Squamata
Flaviviridae	*Flavirus*	Testudines, Squamata , Crocodylia
Paramyxoviridae	*Ferlavirus*	Testudines, Squamata
Retroviridae	*Gammaretrovirus*	Squamata
	Unassigned	Testudines, Squamata, Crocodylia
Rhabdoviridae	Unassigned	Testudines, Squamata
Sunviridae	*Sunshinevirus*	Squamata
Togaviridae	*Alphavirus*	Testudines, Squamata , Crocodylia
Non-enveloped viruses		
dsDNA		
Adenoviridae	*Atadenovirus*	Testudines, Squamata
	Siadenovirus	Testudines
	"Testadenovirus"	Testudines
	Unclassified	Crocodylia
Papillomaviridae	*Dyozetapapillomavirus*	Testudines
	Unclassified	Testudines, Squamata
ssDNA		
Circoviridae	Unclassified	Testudines, Squamata
Parvoviridae	*Dependoparvovirus*	Squamata
dsRNA		
Reoviridae	*Orthoreovirus*	Testudines, Squamata
ssRNA		
Caliciviridae	*Vesivirus*	Squamata
Picornaviridae	"Topivirus"	Testudines
	"Rafivirus"	Testudines
	Unclassified	Squamata

[a]Members of the family *Iridoviridae* can be enveloped but do not generally require an envelope to be infective, although the presence of an envelope may affect their infectivity.

Disclosure Statement: Dr. Marschang is a paid consultant for LABOKLIN, Bad Kissingen, Germany.

References

1. Gaio C, Rossi T, Villa R, et al. Acyclovir administration in marginated tortoise (*Testudo marginata*): pharmacokinetics after single oral administration and considerations about dosage regimen. *J Herpetol Med Surg.* 2007;17:8-11.

2. Allender MC, Mitchell MA, Yarborough J, et al. Pharmacokinetics of a single oral dose of acyclovir and valacyclovir in North American box turtles (*Terrapene* sp.). *J Vet Pharmacol Therap.* 2013;36:205-208.

3. Cooper JE, Gschmeissner S, Bone RD. Herpes-like virus particles in necrotic stomatitis of tortoises. *Vet Rec.* 1988;123:554.

4. Ferguson SD, Waltzek TB, Johnson AJ, et al. In vitro efficacy of cidofovir and acyclovir aganst frog virus 3. *Proc Annu Conf Assoc Reptilian Amphib Vet.* 2014;38.

5. Marschang RE, Milde K, Bellavista M. Virus isolation and vaccination of Mediterranean tortoises against a chelonid herpesvirus in a chronically infected population in Italy. *Deutsche Tierärztliche Wochenschrift.* 2001;108:376-379.

6. Jacobson ER, Gaskin JM, Flanagan JP, et al. Antibody responses of western diamondback rattlesnakes (*Crotalus atrox*) to inactivated ophidian paramyxovirus vaccines. *J Zoo Wildl Med.* 1991;22:184-190.

7. Horner RF. Poxvirus in farmed Nile crocodiles. *Vet Rec.* 1988;122:459-462.

8. Freedom of Information Summary. F10 brand antiseptic solution (benzalkonium chloride and polyhexamide topical solution) Raptors, pet birds, captive small mammals, captive reptiles, and captive exotic/zoo mammals, February 2, 2015. Available at: http://www.fda.gov/downloads/AnimalVeterinary/DevelopmentApprovalProcess/MinMinorUseMinorSpec/UCM440013.pdf Accessed May 17, 2016.

Basic to Advanced Diagnostic Imaging in Reptiles

Nicola Di Girolamo, DMV, MSc(EBHC), GPCert(ExAP), PhD, DipECZM (Herpetology), and
Christoph Mans, Dr vet, med, Dipl ACZM

Session #2134

Affiliation: From the Tai Wai Small Animal and Exotic Hospital, 69-75 Chik Shun Street, Tai Wai, Sha Tin, New Territories, Hong Kong (Di Girolamo), Departments of Surgical Sciences, University of Wisconsin-Madison, School of Veterinary Medicine, 2015 Linden Drive, Madison, WI 53706 USA (Mans).

Abstract: Reptile diagnostic imaging is a field in constant development. Many anatomic and physiologic differences make diagnostic imaging studies in reptiles more challenging to perform and interpret. A proper radiographic study for evaluation of the coelom in chelonians is composed of: (1) a lateral radiograph obtained with a horizontal beam, (2) a cranio-caudal radiograph, also obtained with a horizontal beam, and (3) a dorso-ventral radiograph obtained with a vertical beam. Gastrointestinal contrast radiography is a valuable diagnostic technique, particularly in chelonians and lizards. In particular, in chelonian patients, computed tomography is considered superior to radiography as well as to coelomic ultrasonography, because it overcomes superimposition of structures.

Introduction

Reptile diagnostic imaging is a field in constant development. Many anatomic and physiologic differences make diagnostic imaging studies in reptiles more challenging to perform and interpret.

Radiography

Chelonians

Physical examination of chelonians is greatly limited by their morphology, and diagnostic imaging often is required in order to shorten the list of differential diagnoses, or in order to confirm a clinical suspicion. A proper radiographic study for evaluation of the coelom in chelonians is composed of: (1) a lateral radiograph obtained with a horizontal beam, (2) a cranio-caudal radiograph, also obtained with a horizontal beam, and (3) a dorso-ventral radiograph obtained with a vertical beam. Lateral and cranio-caudal radiographs obtained with a vertical beam are inadequate for assessment of chelonian coelom. It has been demonstrated that positioning chelonians in lateral and in caudal recumbency in order to obtain the lateral and cranio-caudal projections with the vertical beam results in displacement of coelomic organs.[1] In left lateral recumbency the right lung is significantly decreased in volume, and vice versa. The same process occurrs with caudal recumbency, where the caudal part of the lungs decreases in volume. The total volume of the lung does not change with the positioning, due to compensatory increases in nondependent lung tissue.[1]

Another trick to obtain a radiograph that provides the most clinical information is to extend the neck and extremities. Having the head, neck and extremities withdrawn into the shell results in approximately 20% reduction of total lung volume.[1] Extension of the neck and extremities may be obtained without anesthesia. Usually, positioning the chelonian on a cylindrical shape that is small enough in order to prevent the chelonian from touching it with the limbs, results in the extension of the extremities after some time. Stressful procedures (eg, blood sampling, or clinical manipulation) performed before taking the radiographs reduce the chance of success of this technique.

The lungs are best evaluated with a combination of the three projections. The horizontal lateral projection permits the evaluation of the size of the lungs in relationship with the size of the other coelomic organs. In general, a lung height:carapace height ratio of 1:2 is normal. A decrease in lung size may indicate lung disorders or more commonly, enlargement of coelomic organs. A common cause for this finding is the overdistension of the urinary bladder.[2] An increase in lung size may indicate pulmonary emphysema, other primary lung disorders or anorexia with decrease in size of the gastrointestinal tract; however, the horizontal beam lateral projection does not allow diagnosis of the affected lung. The horizontal cranio-caudal projection instead permits screening each lung individually for alterations. Finally, the vertical dorso-ventral projection permits evaluation of each lung field individually but is limited by the superimposition with all the other coelomic organs and shell.

Snakes

In snakes, radiographs can be obtained without anesthesia by use of radiotransparent envelopes or manual restraint in most cases. It is critical to obtain radiographs of snakes once they are completely stretched. It has been demonstrated that imaging the lungs of curled snakes reduces the overall size (width × height) of the central lumen of the lungs as compared to straight snakes.[3] This could result in a more difficult interpretation of pulmonary patterns in curled snakes. Radiotransparent tubes are a common method employed to restrain snakes in a straight position during radiographic studies. The cons of this technique are: (1) it may be stressful to force the snake to exit from the tube, and such stress may not be ideal in a critical patient and (2) snakes in the tube are never perfectly straight and this limits the interpretation of the radiographs.

Two projections (dorso-ventral and lateral) are always required when performing a radiographic study of a snake. The lateral projection usually provides the most information, including evaluation of heart, lungs, gastrointestinal tract, reproductive tract (presence of follicles or eggs), ribs and vertebrae. The dorso-ventral projection is clinically less useful but provides information on vertebrae and ribs that may not be evident on the lateral projection. Specific projections for the head, including one with the head rotated at 60 degrees on the horizontal plane may be needed for diagnosis of fractures or luxation of the quadrate bone. The heart is easily visualized in lateral radiographs as it is the area of soft tissue opacity just cranial to the lungs. The trachea is dorsal to the heart. In snakes with focal enlargement in the proximal third, lateral radiographs may be employed in order to distinguish between mediastinal masses (eg, as seen in lymphoma) and cardiomegaly. The intestinal tract is usually distinguishable by the presence of gas within it. In females, radiology is useful to identify reproductive disorders, even if to a lesser extent than ultrasound.[4] Before reproduction, radiology allows visualization of reproductive material (eggs and fetuses) retained from the previous season and assessment for the presence of coelomic fat. Large follicles may be visualized by noting displacement of the intestinal tract. In snakes with dystocia, radiography usually provides a rough estimate on the number of eggs present, even if the exact number is better determined with ultrasonography; however, radiography is useful in order to evaluate the morphology of the eggs, as it may show change in calcification, size, or internal texture that may be responsible for dystocia.

Lizards

Like chelonians, in lizards the lateral projection yields maximum clinical information when performed with the horizontal beam. The horizontal lateral projection is accompanied with a dorso-ventral projection obtained with the vertical beam. A combination of these two projections permits evaluation of the lungs, the gastrointestinal tract, the reproductive tract and the heart. In chameleons, the lungs provide an optimal contrast and often permit the visualization of the esophagus with the stomach, heart, liver, intestine, kidneys, testicles (in males) and ovaries (in females). In lizards compressed dorso-ventrally (eg, bearded dragons), most of these organs cannot be evaluated in detail unless they are substantially abnormal.

GI Contrast Radiography

Plain radiographs are often of limited use for evaluation of the gastrointestinal tract in reptiles, unless excessive gas or radiodense foreign material is present. Gastrointestinal contrast radiography is a valuable diagnostic technique, particularly in chelonians and lizards.[5] Indications for contrast radiography studies in reptiles include ruling out gastrointestinal obstruction, evaluation of gastrointestinal organ size, integrity of the gastrointestinal tract following trauma or surgery, and for differentiation between the gastrointestinal tract from other soft tissue organs in the coelom.

Both barium sulfate and iodinated non-ionic contrast media have been evaluated in reptiles. Barium sulfate used in prospective studies has been dosed between 5-25 ml/kg and iodinated contrast material has been dosed between 5-7.7 ml/kg. Both contrast media have advantages as well as disadvantages. The selection of contrast media should be based on the indication, species, and underlying or suspected disorder. The type of contrast media affects gastrointestinal transit time. In red-eared slider turtles *(Trachemys scripta elegans),* iodinated contrast material resulted in at least 9-hour faster transit and emptying times compared to barium sulfate, with the exception of gastric transit times.[6] While iodinated contrast material resulted in quicker radiographic results, its radiographic opacity progressively decreased in the distal small intestine and large intestine of red-eared slider turtles. This made visualization more difficult compared to barium sulfate, which resulted in excellent contrast, with no dilution even in the distal intestinal segments.[6] The potential risk of constipation from solidification of barium sulfate particularly in the large intestine of herbivorous reptiles due to the long transit times is of concern. This is not the case if non-ionic iodine-based solutions are used. Further, barium should never be administered to reptiles in which a gastrointestinal perforation is suspected, because leakage of barium will result in inflammation, fibrosis and granuloma formation. There are several types of iodinated contrast media, including low osmolar (non-ionic monomers or ionic dimers), high osmolar and iso-osmolar agents (non-ionic dimers, typically more expensive). High osmolar agents (monomeric ionic) should be avoided in dehydrated patients.[5]

In order to apply the published gastrointestinal contrast studies in reptiles as a reference for interpretation of results obtained in individual reptile patients, it is critical to adhere to the methods used by each respective study, if transit and emptying times are of concern. In particular, type of contrast medium, concentration and volume of contrast medium, environmental temperature and prior food withdrawal should be followed. In reptiles, the gastrointestinal transit time is dependent on the environmental temperature, nutritional status, underlying systemic disease, season, species, age, and amount of contrast media administered. In particular, low environmental temperatures, below the preferred temperature zone, have been shown to have a negative effect on transit times in green iguanas and Greek tortoises.[5] Attempts should be made to maintain reptiles within their reported preferred temperature zones or to provide the same temperature range used in the study that will be used as a reference for transit and emptying times. In bearded dragons, the time of year also had an effect emptying of barium from the colon.[7] The season should be considered in reptiles that undergo seasonal adaptations in their metabolism and behavior (eg, brumation). Transit and emptying times are dependent on total volume of contrast media administered. Smaller volumes will lead to slower transit and emptying times, in particular of the stomach.[8]

Computed Tomography

Computed tomography (CT) is becoming more easily available and affordable for use in reptile patients. In particular, in chelonian patients CT is considered superior to radiography as well as to coelomic ultrasonography in many instances, because it overcomes superimposition of structures, which is of particular advantage in chelonians due to the presence of the shell.[5]

CT of the coelomic cavity can be performed in conscious animals or under mild sedation, if necessary. CT does allow for detailed evaluation of most coelomic organs, including liver, gallbladder, gastrointestinal tract, as well as urinary, reproductive and respiratory systems, and skeletal system. The limited soft-tissue contrast of CT limits interpretation of certain alterations. The intravenous administration of iodinated contrast media is recommended to enhance soft tissue contrast.[9] For chelonians, the administration of iodinated contrast media (either ionic or non-ionic) at a dosage of 600 mg I/kg and a repeated CT scan 1 minute and possible repetition after 3-10 minutes after contrast media administration has been recommended.[9]

On CT scan, the stomach and intestinal loops are easy to recognize due to the gas content, but can otherwise be difficult to differentiate from surrounding soft tissue unless oral contrast media is administered. Increased amounts of gas or fluid in intestinal loops and dilated small intestinal loops with a diameter of 1/3 or more of the shell height are considered abnormal, and might indicate underlying disease.[9]

Free coelomic fluid can be recognized based on density measurement and can lead to elevation of the liver from the plastron.[9] The liver can be evaluated in size. Extension to the inguinal window of the shell is indicative of hepatomegaly. Of particular interest is the density of the liver. The mean density of Hermann's tortoise has been reported to be 50-70 HU (HU: Hounsfield unit).[9] In a study correlating hepatic density of chelonians on CT scan with histopathology, hepatic parenchyma diagnosed with hepatic lipidosis via CT with a hepatic density of less than 20 HU were strongly correlated with severe hepatic lipidosis on histopathology. Liver density values between 20-30 HU correlated with mild hepatic lipidosis histopathologically.[10] This information may provide prognostic information and help educate owners on the importance of nutritional support particularly in anorexic chelonians and lizards.[5]

References

1. Mans C, Drees R, Sladky K, et al. Effects of body position and extension of the neck and extremities on lung volume measured via computed tomography in red-eared slider turtles (*Trachemys scripta elegans*). *J Am Vet Med Assoc.* 2013;243(8):1190-1196.

2. Di Girolamo N, Selleri P. What's your diagnosis? *J Am Vet Med Assoc.* 2016;249(3):271-273.

3. Hedley J, Eatwell K, Schwarz T. Computed tomography of ball pythons (*Python regius*) in curled recumbency. *Vet Radiol Ultrasound.* 2014;55(4):380-386.

4. Di Girolamo N, Selleri P. Reproductive disorders in snakes. *Vet Clin North Am Exot Anim Pract.* 2017;20(2):391-409.

5. Mans C. Clinical update on diagnosis and management of disorders of the digestive system of reptiles. *J Exot Pet Med.* 2013;22(2):141-162.

6. Long CT, Page RB, Howard AM, et al. Comparison of gastrografin to barium sulfate as a gastrointestinal contrast agent in red-eared slider turtles (*Trachemys scripta elegans*). *Vet Radiol Ultrasound.* 2010;51(1):42-47.

7. Karlo H. Radiographic normal anatomy and gastrointestinale contrast studies in bearded dragons (*Pogona vitticeps*). Department fuer Kleintiere und Pferde. Vienna, Austria: Veterinaermedizinische Universitaet Wien;2009.

8. Smith D, Dobson H, Spence E. Gastrointestinal studies in the green iguana: technique and reference values. *Vet Radiol Ultrasound.* 2001;42(6):515-520.

9. Gumpernberger M. Chelonians. In: Schwarz T, Saunders J, eds. *Veterinary Computed Tomography.* Chichester, West Sussex: Wiley-Blackwell;2011.

10. Gumpernberger M, et al. Use of computed tomography to diagnose hepatic lipidosis in reptiles. *Proc Annu Conf Assoc Reptilian Amphib Vet.* 2011;173-174.

Structures and Functions of the Caudal Coelom in Reptiles

Jeanette Wyneken, PhD

Session #2143

Affiliation: From the Department of Biological Sciences, Florida Atlantic University, 777 Glades Road, Boca Raton, FL 33431 USA.

Abstract: Structures and functions of the caudal coelom in reptiles are discussed here. The small and large intestine, liver, pancreas, spleen, some of the hemolymph system (eg, hepatic portal vein, abdominal vein, dorsal aorta and its branches, mesenteric lymphatics), gonads and ducts, and cloaca are contained within the caudal coelom.

Introduction

The reptilian caudal coelom is defined here as the part of the coelomic cavity that is demarked by the pericardial cavity cranially and ventrally, and the cloaca at its caudal-most extent, its major structures are located distal and caudal to the stomach. Because no reptilian species has a diaphragm partitioning the coelom, the designations of what is contained within the caudal coelom are structural and functional. The small and large intestine, liver, pancreas, spleen, some of the hemolymph system (eg, hepatic portal vein, abdominal vein, dorsal aorta and its branches, mesenteric lymphatics), gonads and ducts, and cloaca are contained within the caudal coelom. The surrounding skeleton includes vertebrae, ribs, and in the case of chelonians, much of the plastron.

Skeleton

The trunk vertebrae, ribs, gastralia, and pelvic girdle (or the caudal half of the carapace and plastron of chelonians) are the skeletal elements surrounding the caudal coelom. In the Squamata (snakes and lizards), Sphenodontia (tuataras), Crocodylomorpha (alligators, crocodiles and their relatives), the caudal coelom is bounded by roughly the caudal half of the trunk vertebrae, (sometimes termed the lumbar vertebrae) and their ribs, sacral vertebrae, gastralia (except in snakes) and at least the pubis in some lizards. The posterior-most ribs of tuataras and some lizards are floating ribs that do not attach to the sternum[1]. The Chelonia (turtles and tortoises) vary greatly as to how far the cranially the caudal coelom and its organs extend. Consequently, carapacial landmarks tend to be less useful; instead plastral landmarks usually of more reliable. The caudal coelom of most chelonians usually extends from the posterior margin of the pectoral scutes to ventrally to the caudal aspect of the anal scutes. In some side-neck (Pleurodire) species, the pubis is anklosed to the plastron so may act as part of caudal coelom's skeleton. Chelonian ribs are modified as flattened pleural bones and that articulate with vertebrae forming the carapace[3]. The trunk vertebrae of Chelonia are formed of neural bones that overlie vertebral bodies. The skeleton, muscles and connective tissue bounding the coelom are lined by a coelomic membrane also termed the parietal peritoneum.[1]

Divisions of the Coelom

The reptilian coelom is partitioned by the transverse septum into the pleuroperitoneal cavity and pericardial cavity. The heart and with the bases of the great vessels are surrounded by the pericardial sac, which is part of the transverse septum.[1,3] The pleuroperitoneal cavity extends to the pericardial sac ventrally and cranially; its more caudal extent is to the cloaca (Fig 1). Visceral organs, except for the kidneys, which are retroperitoneal, are located within the pleuroperitoneal cavity. Pleural cavities form secondarily in chelonians, crocodilians and some lizards via a nonmuscular thin fold of tissue that grows from the median mesentery supporting the gut to the peritoneum.[3] The coelomic membrane is the mesothelium, which is formed of is squamous epithelium. It may be darkly pigmented in some squamate species including some agamids, chameleoniids and iguanine lizards,[10] and in the tuatara.[4]

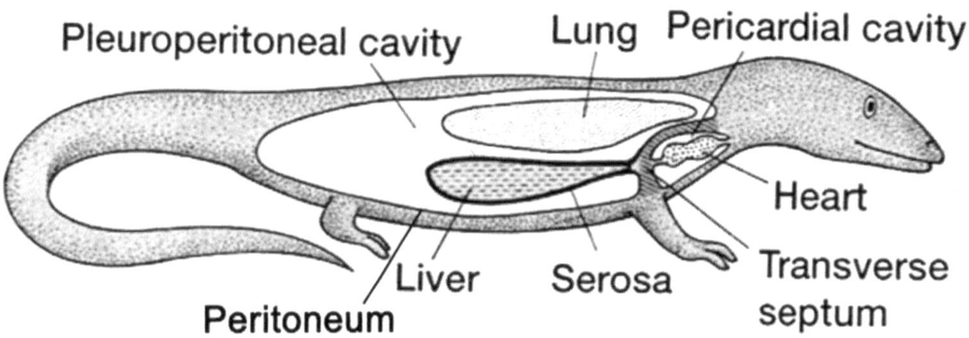

Figure 1. Body cavities of a generalized reptile. The serosa of the liver is formed from the posterior wall of the transverse septum. The coelom is the pleuroperitoneal cavity. A coelomic membrane lines it. Modified with permission from Kardong, KV. *Vertebrates: Comparative Anatomy, Function, and Evolution*. 7th ed. New York, NY: McGraw-Hill;2014.

Liver

The liver is the largest visceral organ in reptiles. It fills much of the pleuroperitoneal cavity, ventral to the lungs, with left and right lobes that may or may not join one another via a bridge of hepatic tissue. The liver attaches to the caudal pericardium via a short coronary ligament (homologous to the falciform ligament of mammals). It is common for the liver to have much of its mass located in the more cranial half of the body, but the caudal margins can extend far caudally in the coelom.

The shape of the liver is variable and generally conforms to the shape of the coelomic cavity. It usually consists of two lobes (except in snakes, which have a single lobe). The right lobe often is larger and attaches to the right lung via hepatopulomonary ligament and the left lobe attaches to the stomach. In chelonians, the liver attaches to the transverse septum via the coronary ligament medially and the left lobe attached to the stomachs' dorsomedial surface by the gastrohepatic ligament. In snakes, the liver lies lateral to the right lung in contact with it.[3,5,6] The liver is caudal to the heart and mostly cranial to the stomach. Across lizard taxa, the liver varies in shape. In some species, it is located primarily on the right side. Iguanine lizards have a liver that can span the coelomic width; some have a well-developed middle lobe. The right lobe often has an elongated tail that extends caudally and encases part of the postcava. The liver of the tuatara has thin left and right lobes and a large dorsomedial lobe; it is reported to have ligamentous attachment to the pelvis.[4] The liver of crocodilians is large and the right lobe tends to be larger than the left. The liver is lies cranial and ventral to the posthepatic septum (also termed a pseudodiaphram), which separates the pleural, cardiac and hepatic cavities from the more caudal visceral coelom.[3,7] The liver, post-hepatic fat body, and visceral mass together move cranially and act a hepatic piston to function in lung compression during exhalation.[8]

The liver's main functions are fat metabolism, fat storage, protein metabolism, glucose supply, filtration and production of exocrine products. Fats entering the liver from the gut are bound proteins and converted to phospholipids and cholesterols, which function structurally and in the formation of bile salts[3]. Bile aids in digestion and absorption. Bile travels to the duodenum via the common bile duct, which is lined by cuboidal epithelium and goblets cells.[8] Bile is transported to and from the gallbladder via the cystic duct. Most taxa have gall bladders. The gallbladder located in the right lobe of the liver in chelonians and lizards. It is caudal to the liver in snakes.[5,6,9] *Hydrosaurus pustolosus* (sailfin water dragon) is reported to lack a gallbladder and bile drains directly to the duodenum.[11]

The liver also functions in excretion of cholesterols and the breakdown of hemoglobin.[12] Carbohydrate metabolism leads to glucose formation in hepatocytes as glycogen or via pyruvate or lactate. Under anaerobic conditions, the liver can produce lactate. The liver plays a major role in hibernation. It functions in blood sugar regulation and protein metabolism. Several forms of the hormones known as clotting factors (fibrinogen, prothombrin) are formed in the liver.[12] Vitellogenins, which are responsible for yolk formation, are formed by the liver.[12]

Pancreas

The pancreas is a lobular organ in the mesentery supporting the proximal duodenum in lizards, crocodilians, turtles and the tuatara. In snakes, it may be a single oval organ or form with the spleen as a splenopancreas. The pancreas is both an endocrine and exocrine organ. Pancreatic cells produce two major kinds of exocrine products: alkaline fluid and digestive enzymes (proteolytic enzymes, amylases, and lipases). These products enter the duodenum via the pancreatic duct that opens into the ampulla of Vater.[3,7] The pancreatic duct joins the common duct in colubrid snakes. In some iguanine lizards, the common bile duct enters the ampulla of Vater directly while other squamates have paired bile ducts that each enters the duodenum separately from the pancreatic duct. The endocrine role of the pancreatic islet cells is the production of insulin and glucagon, which regulate blood glucose.[3,13-15]

Intestines

The intestines function in digestion and absorption. Intestine length, along with texture of the mucosa, contribute to the surface area for digestion. As digesta passes through the intestines, the mucosa adds secretions, absorbs nutrients and water, and ions. The combined actions of longitudinal and circumferential muscle move digesta and ensure that surface area is thoroughly exposed for absorption. The intestines are suspended from the dorsal body wall by the large mesentery. The mesentery, formed of two layers of peritoneum, supports many branches of the superior and inferior mesenteric arteries and veins, and lymphatic vessels that transport the products of digestion as well as nerves.

The small intestine may be either straight or be composed of short transverse loops. In snakes, the intestine may have many largely longitudinal folds that allow for expansion and high surface area during digestion. Organ size is dynamic and responsive. In garter snakes (*Thamnophis sirtalis parietalis*) that were fed frequently, liver and intestine size was large while in snakes fed less frequently there was a marked up- and down regulation of organ size.[16] The small intestine in many lizard species and turtles has many loops and convolutions, while that of snakes is relatively simple with alternating bends. The length of the small intestine can vary greatly with species. In general carnivorous species tend to have the shortest small intestine length and herbivores have the longest.[15] The small intestine of the tuatara is short, as is the colon.[4]

The small intestine ends in an ileocecal valve or sphincter. The caecum and proximal large intestine tends to be larger in herbivorous and omnivorous species than in carnivores. A caecum is present in most lizards and chelonians, but absent in most snakes, except for pythons. Crocodilians appear the lack a caecum.[3] In some herbivorous lizards, including members of the Agamidae, Scincidae, and Iguanidae, the distal ileum to proximal colon is partially compartmentalized. In *Iguana iguana*, the cecum and proximal colon have folds that may function as valves.[14,17,18]

The large intestine is usually pink in color; however, it may be darkly pigmented (in many chameleons and some turtles). It tends to have a much larger capacity in herbivorous species than in carnivores and omnivores.[19] Large intestine length is longest in herbivores and shortest in carnivores.[15] The large intestine functions in reclaiming water, absorbing calcium ions, and degrading complex carbohydrates.

The food transit times through reptilian GI systems are variable and are usually measured in days to weeks. In addition to thermoregulatory behavior, several mechanisms influence transit times and digestive efficiency. The digestive efficiency of reptile intestines is much lower than that of mammals, largely because mammals have much greater surface area and to some extent because reptile digestive efficiency is linked to basking or other post-feeding thermal regulatory behavior.[12,15,20] Digestion in some herbivorous species is aided by fermentation and in some species by the activity of some nematodes. In some herbivorous chelonians and lizards, microbial fermentation occurs in the large intestine.[19] Fermentation has been documented in the small intestine of one species of freshwater turtle *Pseudemys nelson*.[19] For example, in species that ferment digesta in the hindgut, antiperistalsis may increase transit time and allow functional flexibility how different foods are transported through the gut.[19]

Urea and/or uric acid transfer into the intestine by retrograde movement of urine into the intestine in reptiles. Intestinal microbes in the reptile colon appear to degrade urea and uric acid. While nitrogen assimilation levels are much lower in reptiles than in birds and mammals, nitrogen reclaimed in the colon is co-opted for amino acid synthesis.[12,17]

Vasculature

The paired ventral abdominal veins (in the Chelonia) or a single abdominal vein (in the Squamata, Sphenodontia, and Crocodylomorpha) travels through the peritoneum and passes into the liver through the coronary ligament.[13,21,22] The major vessels passing from the caudal coelom to the cranial coelom include (but are not limited to) the dorsal aorta, superior mesenteric artery, ascending vena cava (synonym: post cava), renal arteries and veins, gonadal arteries and veins, hepatic portal vein, ventral abdominal veins, and epigastric arteries to the fat bodies.

Cloaca

The large intestine terminates at the anus that opens into the cloaca. The cloaca is the exit point for organs of the caudal coelom but is not part of the coelom. The cloaca is a short, large-bore tube. The reptilian cloaca is often described the regional openings based upon the body's products they receive. Their morphological distinctness from one another varies across taxa. The feces are discharged into the most craniad chamber, the coprodeum. The urodeum receives nitrogenous wastes and gametes via the urogenital ducts. The largest chamber is the posterior proctodeum acts as a general collecting area for digestive and excretory wastes.[3,7] Often these regions are in such close proximity to one another that describing them as unique chambers is not informative.

References

1. Hyman LH. *A Laboratory Manual for Comparative Vertebrate Anatomy.* Chicago, IL: University of Chicago Press;1922.

2. Pritchard PCH. Evolution and structure of the turtle shell. In: Wyneken J, Godfrey MH, Bels V, eds. *Biology of Turtles: From Structures to Strategies of Life.* Boca Raton, FL: CRC Press;2008:45-84.

3. Kardong, KV. *Vertebrates: Comparative Anatomy, Function, and Evolution.* 7th ed. New York, NY: McGraw-Hill;2014.

4. Newman AK. Notes on the anatomy and physiology of the tuatara. Trans NZ Inst II. 1877;17:22-239.

5. Lawson R. The comparative anatomy of the coelom and the digestive and respiratory systems. In: Wake MH, ed. *Hyman's Comparative Vertebrate Anatomy.* Chicago, IL: University of Chicago Press; 1979: 378-447.

6. Schaffner F. The liver. In: Gans C, Gaunt AS, eds. *Biology of the Reptilia.* Ithaca, NY: Society for the Study of Amphibians and Reptiles. 1998;19:485-531.

7. Romer AS, Parsons TS. *The Vertebrate Body.* Philadelphia, PA: Saunders; 1986.

8. Claessens LPAM. A cineradiographic study of lung ventilation in *Alligator mississippiensis*. *J Exp Zool* A. 2009;311A:563-585.

9. Luppa H. Histology of the digestive tract. In: Gans C, ed. *Biology of the Reptilia.* London, UK: Academic Press;1977:225-314.

10. Krejsa RJ. The comparative anatomy of the integumental skeleton. In: Wake MH, ed. *Hyman's Comparative Vertebrate Anatomy.* Chicago, IL: University of Chicago Press;1979:112-191.

11. Auffenberg W. *Gray's Monitor Lizard.* Gainesville, FL: University Press of Florida;1988.

12. Skoczylas R. Physiology of the digestive tract. In: Gans C, Gans KA, eds. *Biology of the Reptilia.* London, UK: Academic Press;1978:589-717.

13. Divers SJ. The liver and gastrointestinal testing. In: Fudge AM, ed. *Laboratory Medicine: Avian and Exotic Pets.* Philadelphia, PA; Saunders;1999:205-209.

14. Jacobson E. *Infectious Diseases and Pathology of Reptiles: Color Atlas and Text.* Boca Raton, FL: CRC Press;2007.

15. Stevens CE, Hume ID. The digestive system of fish, amphibians, reptiles, and birds. In: Stevens CE, Hume ID, eds. *Comparative Physiology of the Vertebrate Digestive System.* New York, NY: Cambridge University Press;2004:24-45.

16. Starck JM, Beese K. Structural flexibility of the small intestine and liver of garter snakes in response to feeding and fasting. *J Exp Biol.* 2002;205:1377-1388.

17. Mader D. *Reptile Medicine and Surgery.* St. Louis, MO: Elsevier Saunders;2006.

18. Parson TS, Cameron JE. Internal relief of the digestive tract. In: Gans C, ed. *Biology of the Reptilia.* London, UK: Academic Press;1977:159-223.

19. Bjorndal KA. Fermentation in reptiles and amphibians. In: Mackie RI, White BS, Isaacson RE, eds. *Gastrointestinal Microbiology: Gastrointestinal Ecosystems and Fermentations.* Vol. 2. New York, NY: Chapman and Hall;1997:199-230.

20. Harlow HJ, Hillman SS, Hoffman M. The effect of temperature on digestive efficiency in the herbivorous lizard, *Dipsosaurus dorsalis*. *J Comp Physiol.* 1976;111:1-6.

21. Greene H. Feeding in snakes. In: Schwenck C, ed. *Feeding: Form, Function, and Evolution in Tetrapod Vertebrates.* San Diego, CA: Academic Press;2000:293-333.

22. Wyneken J. *The Anatomy of Sea Turtles.* Miami, FL: US Department of Commerce NOAA Technical Memorandum NMFS-SEFSC-470;2001.

Pathology of the Reptilian Caudal Coelom Focusing on the Urogenital and Hepatobiliary Systems

Drury R. Reavill, DVM, DABVP (Avian Practice), DABVP (Reptile and Amphibian Practice), DACVP

Session #2052

Affiliation: From Zoo/Exotic Pathology Service, 6020 Rutland Drive #14, Carmichael, CA 95608, USA.

Abstract: This paper will cover disease conditions of the reptilian urogenital tract and hepatobiliary system, including the pancreas and adrenal glands. It is not a comprehensive review. More in-depth disease descriptions can be found in the articles, website, and books in the reference list.

Introduction

Common insults to the reptilian urogenital and hepatobiliary systems include developmental anomalies, degenerative lesions, obstructions, traumatic injuries, infectious disease agents, and neoplasms. The paper is divided by location (kidney, urinary bladder, reproductive system, liver/gall bladder, pancreas, adrenal gland) and disease categories (congenital, nutritional, metabolic, toxic, aging, idiopathic, infectious-viral, bacterial, mycotic, parasitic, and tumors). A short discussion of these disease conditions in reptiles is found in each section. For the purposes of this paper, Chelonia will refer to all of the shelled reptiles. The term tortoise will be used to describe the large land animals that have stout elephantine feet. Few references to marine turtles and crocodilian species are included, as these are uncommonly seen in most clinical practices.

Kidney

Congenital abnormalities

There are a number of described congenital abnormalities of the kidneys. Rarely agenesis of parts or a whole kidney has been reported in snakes as well as an extra kidney in an anaconda (*Eunectus murinus*).[1]

Renal cysts are characterized by variable cystic formations within the renal parenchyma.[1] Some cysts are derived from greatly dilated renal tubules. Glomeruli can also develop dilated cystic Bowman spaces. The cysts may be identified throughout one or both kidneys. Some cases of renal cyst development may be due to underlying chronic inflammatory diseases. Tubules may become markedly dilated if there is interstitial fibrosis and inflammation. Gout tophi may also form in these cases.[2] There has been a reported case of a renal and hepatic cyst development in a giant tegu (*Tupinambis merianae*). This congenital condition (polycystic disease) has been reported in mammalian species.[1]

Nutritional disease

Hypovitaminosis A can lead to squamous metaplasia of mucosal epithelium. This metaplasia can involve the renal collecting ducts, and the accumulation of desquamated cornified cells may result in a partial blockage. This can result in the formation of cystic spaces.[1] Concurrent and/or subsequent uric acid deposition also has been described in crocodiles.[3] This metaplasia and hyperkeratosis of the renal tubules will decrease the renal clearance of the urates. With a hyperuricemia, there will be a precipitation of uric acid crystals and the development of gout tophi in multiple organs.

Metabolic disease

Amyloid deposition has been poorly documented, but there are small numbers of reports in reptiles. These describe deposition of material consistent with amyloid in a Central American boa (*Boa constrictor imperator*) and a rainbow water snake (*Enhydris enhydris*).[4] An amyloid-type material has also been identified in the renal interstitium and mesangium of the glomeruli in four komodo dragons (*Varanus komodoensis*).[5] Amyloid is an insoluble pathologic proteinaceous substance, deposited between cells in various tissues and organs of the body. This amorphous, eosinophilic hyaline extracellular substance encroaches on and results in pressure atrophy of adjacent cells. Special stains and electron microscopy can be helpful in further classifying these rare deposits.

Nephrocalcinosis is due to the deposit of calcium salts (mineral) into the soft tissues and is a common findings in reptiles.[1,6,7] This deposition may be identified within the glomerular mesangium or basement membranes in both the glomeruli as well as the renal tubules. Calcium deposits may be recognized within the interstitium as well as involving the vascular structures. There are two general mechanisms of tissue mineralization: dystropic and metastatic. Dystrophic calcification occurs in nonviable or dying tissues. Dystrophic is not associated with hypercalcemia or other disturbances of calcium homeostasis. It occurs in cells injured in a variety of ways, including vascular, toxic, metabolic, or inflammatory diseases. Metastatic calcification results in the deposition of calcium in vital tissues and reflects some disturbance in calcium metabolism. In renal disease, it is difficult to differentiate dystrophic (degenerative lesions of uremia) from metastatic calcification (secondary hyperparathyroidism). Histologically calcification is characterized by irregular foci of dense basophilic (blue) material that can be identified by special stains. Additional sites of mineral deposition include the great vessels, heart, lung, intestines and skin.[8,9]

Pigment deposition identified as granular golden-yellow to brownish material are not uncommonly identified in the cytoplasm of the renal tubular epithelial cells. The identity of this pigment is not definitively determined; however, one reference suggests that it is lipofuscin.[1] Lipofuscin is an intralysosomal pigment formed by peroxidation and polymerization of unsaturated fatty acids. While not injurious to the cell, this pigment is a sign of free radical injury and lipid peroxidation. This occurs with loss of antioxidant defense mechanisms such as hypovitaminosis E. It is also associated with aging, severe malnutrition, and cancer cachexia. Other materials such as hemosiderin or bile pigments are possible and special stains can be helpful in determining the classification. Inexperienced pathologists could possibly confuse these intracytoplasmic pigments with the cytoplasmic inclusions of inclusion body disease (IBD), which is discussed below.

Xanthomas (cholesterol-rich granulomas) are reported in species of geckos. A deposition of cholesterol crystals and associated inflammation has been described in round island geckos (*Phelsuma geuntheri*).[10] In these geckos, the primary lesion was of a lipidosis involving the liver as well as the spleen. The cholesterol crystals are acicular clefts associated with chronic and acute inflammation (xanthomas) and identified in the kidney and liver. In another report of different gecko species, these systemic xanthomatous changes were not identified in the kidney.[11]

Aging

The majority of reptiles that have renal disease generally have chronic changes in which the etiology can no longer be determined. These chronic changes include variable degeneration and/or necrosis of glomeruli and/or tubules. Interstitial and glomerular fibrosis may be prominent. There can be variable dilation of tubules and, in some cases, calcification as well as gout tophus formations. These types of lesions are more common in the common green iguana and other large lizards, and are suspected to be of multifactorial etiology, including poor nutrition such as extremes in protein, chronic dehydration, and imbalanced vitamin and/or mineral supplements.[1]

Toxins

Gentamicin has been demonstrated to be nephrotoxic, particularly in snakes.[12] The lesions progress from a cloudy swelling of the proximal tubules to hydropic degeneration and finally to tubular necrosis. The snakes in the high-dose group developed visceral gout and also had extensive tubular necrosis. The tophi were identified on the pericardium, the serosal membranes and parenchyma of the kidney, liver, spleen, and lungs. A similar lesion has also been described in snakes and a South African monitor (*Varanus exanthematicus*) that were exposed to the insecticide azinphos-methyl.[1]

Renal gout is a deposition of uric acid in the soft tissues; another common lesion in reptiles.[1,3,6,13-16] There is a variable percentage of uric acid that comprises the end product of protein excretion from the body in both reptiles and amphibians. In general, the percentage excreted as uric acids is most likely related to the animal's natural habitat (eg, desert tortoises [*Gopherus agassizii*] convert wastes to the insoluble salts of uric acid).[17] The cause of tissue urate deposition is renal dysfunction. Early gout lesions will have uric acid crystals deposited within the tubular epithelial cells with gradually more crystals accumulating around these centers. As the disease progresses, these will break through the basement membrane of the tubule, forming the typical gout tophi. The gout tophi has radiating uric acid crystals surrounded by variable numbers of inflammatory cells. In more acute cases, these inflammatory cells may be degenerate heterophils and, as the lesions become more chronic, there will be macrophages and multinucleate giant cells. Gout tophi are most commonly identified associated with the tubules, although they may also be found within the interstitium as well as within the glomerular mesangium.

It is suspected that gout may be induced by dietary factors. Such factors can be an inappropriate diet fed to herbivorous reptiles (eg, green iguanas fed high-protein dog and cat foods). In crocodiles, gout was associated with suspected hypovitaminosis A.[3] There has also been a suggestion that chronic low-grade dehydration can impair renal function and lead to gout.[16] Renal toxins are linked to gout formation within the kidney.[12]

Infectious disease: viral

IBD of boid snakes can be histologically recognized in lining mucosal epithelial and lymphoid cells aggregates of the esophagus, as well as within the mucosal cells of the stomach and less frequently in the intestines.[18] The eosinophilic intracytoplasmic inclusions have been described in boas and pythons (family Boidae), palm vipers (*Bothriechis marchi*) and corn snakes (*Elaphe guttata*).[18-20] It is important to note that these inclusions do not always confirm IBD, as the inclusions in the corn snakes did not have consistent morphology when examined by electron microscopic.[19] This disease can be responsible for clinical signs including chronic regurgitation, incoordination, loss of righting reflexes, paresis, and an increased incidence of secondary infections such as stomatitis and pneumonia. Clinical signs referable to the central nervous system are more prominent in the Pythoninae subfamily.[21,22] Antemortem diagnostics include CBC's and biopsies of the esophagus (especially of lymphocytic stromal aggregates or "tonsils"), gastric mucosa, and liver. Post mortem samples should include the kidney, pancreas, brain (especially in pythons) and the liver. The etiologic agent of IBD is an arenavirus.[23]

Infectious disease: bacterial

Any number of bacterial infections can involve the urinary system. The majority of these in both reptiles and amphibians are gram-negative bacteria and primarily *Pseudomonas* or *Aeromonas* species. With bacterial septicemias, the renal blood vessels within the glomerular capillaries as well as in the interstitium may be blocked by clusters of the bacterial organisms. There may also be a peripheral heterophilia in response to a bacteremia.

Mycobacterial infections, which are generally systemic diseases, can be identified within the kidney. Although all reptiles are suspected to be susceptible, the reports are rare. Those with renal involvement include a Kemp's ridley sea turtle (*Lepidochelys kempii*) with systemic *Mycobacterium chelonae* and disseminated mycobacteriosis in freshwater crocodiles (*Crocodylus johnstoni*).[24,25] The granulomas typically have central necrosis supporting large numbers of acid-fast rod-shaped bacteria and surrounded by multinucleate giant cells as well as macrophages. There may be a fibrotic capsule, depending on the chronicity of the lesions.

Infectious disease: mycotic

Chromomycosis is primarily a systemic fungal infection by pigmented fungi of the family Dematiaceae.[26] These fungi are saprophytes that live in soil and decaying vegetable matter, predominantly in warmer climates of tropical and subtropical countries.[27] Some identified members include *Cladosporium herbarum*, *Scolecobasidium humicola*, *Fonsecaea* spp., *Phialophora* spp., *Rhinocladiella* spp., *Hormodendrum* spp., *Curvularia* spp., and *Drechslera* spp.[28] The fungal organisms produce a characteristic amber-brown, thick-walled, septate structures known as sclerotic bodies.[28] Chromomycosis infection is most commonly described in amphibians but has also been reported in mammals, a radiated tortoise, pythons, boas, eastern box turtle (*Terrapene carolina carolina*), and a mangrove snake with a concomitant fibrosarcoma.[29,30] Transmission is suspected to be by the alimentary tract and/or skin abrasions. Skin trauma incurred during feeding may be the most likely portal of entry (especially in colonies of animals or with vigorous eaters). The stress of captivity or other debilitating disease processes can also contribute to the development of the disease.

Chromomycosis generally presents first as an ulcerated or granulomatous skin lesion. These lesions can be nodules, grey-black papules and/or irregularly shaped grey-black ulcers. The most common sites for external lesions are the dorsal and ventral skin surfaces. Over a period of time (which can be greater than six months) the animal succumbs to disseminated disease. On gross examination multiple organs may contain multiple coalescing grey-black nodules. These granulomas are commonly described in the liver and kidney, but also occur in the spleen, lung, heart, adrenal, gastrointestinal tract (serosal and mucosal surfaces), bone, and central nervous system.

Exfoliative cytology of the cutaneous lesions may have mononuclear cells surrounding pigmented, septate fungal hyphae.[26] The fungal organisms are not readily isolated by culture, as they are slow growing and temperature sensitive (25-27 °C).

Histologically the multifocal granulomas are comprised of macrophages, lymphocytes, plasma cells, and multinucleated giant cells. They contain brown-pigmented fungal elements which are spherical, thick-walled sclerotic bodies (8-12 µm in diameter) with equatorial and occasionally vertical septations. Pseudohyphae and, rarely, septate hyphae can be found. The fungal organisms may be found extracellularly as well as within the cytoplasm of multinucleated giant cells. These organisms are zoonotic with human infection resulting from the contaminated environmental.[28]

Infectious disease: parasitic

Metazoan parasites: Trematodes identified as *Styphlodora renalis* and *Styphlodora horrida* have been described in the boa and python.[31-33] Ureters were dilated and filled with the parasites as well as urates, cellular debris, and small amounts of calcium. In the boa constrictor, the trematodes *Styphlodora horrida* were found in the kidney. Trematodes have also been reported in king snakes, cottonmouth, indigo snakes, tropical rat snakes, black stripe snakes, bushmasters, as well as boas.[34]

Spirorchid trematodes (blood flukes) have also been identified in the ureters of painted turtles (*Chrysemys picta*) and red-eared slider turtles (*Trachemys scripta elegans*) which also had ureteroliths.[35,36] The lesions are due to microgranulomas forming around the ova lodged within the blood vessels.

Strongyloids, which have an alternating free moving and parasitic life cycle were recognized in one case of a Burmese python (*Python molurus bivitattus*). These nematodes were identified in the ureters, kidneys, as well as the small intestine. They were associated with a ureteritis as well as an enteritis. The ureters were bilaterally obstructed. It is felt that these Strongyloides had possibly migrated from the cloaca.[37]

Microfilaria have also been reported in the great vessels or at the renal vein of reptiles. These have not been associated with significant pathology.[1]

Protozoan parasites: An intranuclear protozoa is reported in several species of captive tortoises in the United States.[38] This has been described in two captive-bred juvenile radiated tortoises (*Geochelone radiata)*, two adult radiated tortoises, one wild caught adult impressed tortoise (*Manouria impressa*), one captive-bred juvenile leopard tortoise (*Geochelone pardalis*), three Travancore tortoises (*Indotestudo forstenii*) and four young adult radiated tortoises. All tortoises had intranuclear coccidial parasites in a variety of epithelial tissues. Inflammation of the lung was noted in five tortoises and one had a proliferative pneumonia.[38-40]

A similar coccidian has also been described in the collecting ducts and ureters of the boa constrictor. The exact classification is currently unresolved but may be *Klossiella boae* or *Tyzzeria boae*. The infected epithelial cells are enlarged but appear to result in little pathology.[1,41]

Entamoeba invadens can cause high morbidity and mortality among some snakes and chelonians.[42-46] and rarely in lizards.[42] *Entamoeba invadens* is a member of the *Entamoeba histolytica* group of amoebae. This protozoon is an important obligate parasite that has a direct life cycle. It is common to see secondary bacterial infections associated with these organisms. In snakes and chelonians, the infections start primarily within the colon. From here, they will be found, by the portal circulation, in the liver and other organs including the kidney.[1,47,48] The lesions are of extensive necrosis with inflammation of macrophages, heterophils, and lymphocytes. The protozoa are approximately 10-15 µm in diameter, round, and stain poorly on hematoxylin and eosin (H&E) sections.

Microsporidia are small, intracellular protozoans. They form an infective thick-walled spore containing the sporoplasm which once ingested, is discharged and migrates to its target organ. Microsporidian protozoa include *Encephalitozoon* sp., *Pleistophora* sp., *Nosema* sp. and other species. The reports in reptiles are primarily of systemic fatal infections in bearded dragons (*Pogona vitticeps*).[49] The primary lesion is a severe hepatic necrosis with clusters of light basophilic intracytoplasmic microorganisms packed within distended hepatocytes as well as being free in areas of necrosis. Similar microorganisms can be found in cytoplasmic vacuoles of renal epithelial cells, pulmonary epithelial cells, gastric mucosal epithelial cells, enterocytes, capillary endothelial cells, ventricular ependymal cells in the brain, and in macrophages within granulomatous inflammation of the colon, adrenal glands, and ovaries. They are oval, approximately 1 µm in length, and stain faintly on H&E sections. The spores are acid-fast positive and stain variably gram positive.

There are many flagellate protozoa found in reptiles. Most reside in the lumen of the intestines; however, there are flagellates associated with disease found in the urogenital system of reptiles. These are of the genus *Spironucleus* and have been recognized as a cause of disease in chelonians and snakes.[1,50] Cases in chelonians are subacute and the infections can be persistent. Tubulointerstitial nephritis and necrosis are the common renal lesions with the protozoa accumulating within the lumen of the tubules. The majority of cases seen by the authors have been in box turtles (genus and species not provided), a water dragon, and geckos (species not provided). The life cycle is direct with transmission by ingestion of food contaminated with affected urine or feces. They can be identified on fresh urine or fecal direct smears. The trophozoites are pyriform in shape and have multiple flagella.

Myxosporidia (protozoa of the class Myxosporea and phylum Myxozoa) are common parasites of fish, and uncommon parasites of reptiles. The genus *Myxidium* is best described in the urinary system of chelonians; however, they have not typically been associated with significant pathology. Recent reports indicate this protozoa can be a cause of renal disease[1,51-53] in chelonians, primarily aquatic turtles (Indo-Gangetic flap-shelled turtles [*Lissemys punctata andersonii*], Crowned River turtles [*Hardella thurjii*], and red-eared slider turtles [*Trachemys scripta elegans*]).[54] Histologically, renal intratubular myxozoan spores were associated with renal tubular necrosis, tubular mineralization, peritubular fibrosis and dilation, and chronic interstitial nephritis, with membranoproliferative glomerulopathy. Specific identification of these protozoa requires electron microscopy and genomic analysis.

Renal tumors

The principal renal tumors reported in reptiles are adenocarcinoma, adenoma, and nephroblastoma. These tumors have been described in lizards, snakes, and chelonians.[14,55-58] Several references list tumors affecting the kidney, including the primary renal tumors.[1,59,60] In lizards, renal tumors present as firm swellings in the caudal coelom and can occasionally be palpated extending cranially past the rim of the pelvis. In snakes, swelling in the caudal third of the body cavity is typical. Grossly one or both kidneys may have irregular masses within the parenchyma.

Differential diagnoses for these lesions include acute and chronic renal disease in lizards and impacted eggs in snakes. Occasionally in snakes, renal tumors have been mistaken for impacted eggs and massaged out of the body through the cloaca. Metastasis is unusual but has been reported. Sites include liver, lung, and perirenal abdominal wall. Surgical removal of renal tumors is recommended.

Urinary Bladder

Any microscopic precipitates or polycrystalline concretions found in the urinary tract are classified as urinary calculi. These urinary calculi, uroliths, have been reported in lizards, chelonians, and snakes.[61,62,63] Most are located in the bladder of those animals with a urinary bladder, and uncommonly some cases are described in the ureters and cloaca.[47] For species without urinary bladders, such as snakes, the concretions may be found in the distal ureters where urine is frequently stored. The cause of the calculi formation is unknown. Some proposed etiologies include nutritional deficiencies, such as of vitamin A and D. Excessive dietary protein, oxalates, as well as bacterial infections and suture remnants have been reported as causes. Some of these factors contribute to or form a nidus on which concentric layers of material are laid. A common clinical finding with calculi formation is dehydration. In tortoises, a phenomenon recognized in females is of fully developed eggs being retropulsed into the urinary bladder. They apparently are unable to be passed back out and will develop into a calculus.

The majority of small calculi generally will not result in significant lesions or clinical signs. If there are clinical signs they are of anorexia, constipation, egg binding, dysuria, and poor growth.[61,62] Calculi with rough surfaces or increasing sizes may irritate the lining of the urinary bladder, resulting in hematuria as well as hypertrophy of the bladder wall and hyperplasia of the lining epithelial mucosa. Pressure necrosis to the bladder wall as well as internal viscera have been reported, particularly with large stones.[7,62,64] Uroliths of reptiles and amphibians all appear to be composed of a type of urate salt. This urate salt may be complexed with calcium and phosphates.[63,64] Rarely cystic calculi primarily composed of apatite (calcium phosphate) are reported.[63]

Reproductive System

Congenital abnormalities

Hermaphroditism (having reproductive organs normally associated with both male and female sexes) has been reported in a phenotypic male common green iguana (*Iguana iguana*).[65] An intersex Madagascar tree boa (*Sanzinia madagascariensis*), housed with a male and recorded as breeding but never produced eggs supported the well developed gonads with maturing ovarian follicles and mature sperm within seminiferous tubules.[66] Intersex/hermaphroditism has been described in various papers from the early 1900's in tortoises, lizards, and turtles.[66]

Salpingitis and yolk retention

Inflammation of the oviduct/uterus can be due to infectious agents, or secondary to retained yolk with impaction. The inflammation in the latter case is secondary to irritant tissue effects of the yolk protein.

A study was conducted to compare thiamine concentrations in American alligators (*Alligator mississippiensis*) and Florida largemouth bass (*Micropterus salmoides floridanus*). Results suggest that thiamine deficiency might be playing an important role in alligator embryo survival. These results suggest that thiamine deficiency might be playing an important role in alligator embryo survival but not in reproductive failure and recruitment of largemouth bass. The cause(s) of this thiamine deficiency are unknown but might be related to differences in the nutritional value of prey items across the sites studied and/or to the presence of high concentration of contaminants in eggs.[67]

Tumors

Seminomas are reported in snake, lizard species, and one report in a spur-thighed tortoise (*Testudo graeca*).[68] The seminoma was unilateral and large enough to compress the liver.

Teratomas are composed of tissues derived from at least two of the three embryonic germ layers (endoderm, mesoderm, and ectoderm). Typically, they arise from the gonads. The ovaries are common sites for teratoma development in the common green iguana (*Iguana iguana*)[69] and other unspecified saurian,[59] as well as a chelonian.[70] Clinical signs are the result of a space-occupying mass and may include digestive and respiratory dysfunction as well as coelomic cavity distention. Differentials in females include neoplasia or granuloma of the ovary or oviduct, cystic ovary, segmental pyosalpinx, and neoplasm or granuloma of the adrenal gland or spleen. In the few described cases the tumors had a rapid growth rate and a potential for malignancy. Early surgical removal is recommended.

On the basis of the morphologic and immunohistochemical results, a diagnosis of ovarian undifferentiated carcinoma was made in a corn snake (*Elaphe guttata guttata*). Histologic specimens consisted of a highly cellular mass composed of pleomorphic, spindle-shaped cells and, occasionally, round to polygonal cells arranged in irregular fascicles. The neoplastic cells were immunoreactive for cytokeratin (AE1/AE3), smooth muscle actin, and skeletal muscle actin, but did not stain for vimentin or desmin.[71]

A European viper with a cloaca and oviduct adenocarcinoma died after photodynamic therapy with chloroaluminum sulfonated phthalocyanine.[72]

Liver

Congenital abnormalities

In a spur-thighed tortoise (*Testudo graeca ibera*) the two heads and necks were independent, with duplicated esophagus, stomach, liver, gallbladder, trachea, and two asynchronous hearts.[73] The proposed causes reported are exposure to environmental toxins, genetic defects, anoxia of the embryo, elevated environmental temperatures during incubation, and rapid thermal changes during internal incubation.[73,74]

Infectious disease: viral

Adenoviruses (AdVs) have been found to infect all classes of bony vertebrates and most appear to have co-evolved with their hosts. They are double stranded non-enveloped DNA viruses. Adenoviruses fall in the family Adenoviridae, which is further classified into 5 currently accepted genera: Mastadenovirus isolated from mammals; Aviadenovirus in birds; Atadenovirus isolated from squamates; Siadenovirus found with some chelonians, amphibians and birds; and Ichtadenovirus isolated from a sturgeon. A sixth genus, Testadenovirus, isolated from tortoises, has been proposed. AdV infections have been identified in wild-caught and captive lizards, snakes, chelonians, and crocodiles. Disease often is a combination of other disease agents and/or stress and the viral infection. Adenoviral infections in reptiles have ranged from clinically inapparent infections to lesions of enteritis, necrotizing hepatitis, esophagitis, splenitis, and encephalopathy.

Colubrid snakes can present with coelomic cavity fluid, enlarged and mottled liver, dilated small intestines. On histology there are large basophilic or amphophilic inclusions in mucosal epithelial cells of the stomach, intestines, esophagus and liver. Specifically there is enteritis, hepatic necrosis, secondary bacterial septicemias, foci of gliosis primarily within the caudal cerebrum, brainstem, and cervical spinal cord.[75,76]

The best described infections have been in bearded dragons. Grossly there may be coelomic effusion and hepatic enlargement. Basophilic intranuclear viral inclusions and severe coagulative necrosis of the liver, intestine, lymphohistiocytic interstitial nephritis, and bile duct are commonly described in histology.[77]

Sulawesi tortoises may grossly have hepatosplenomegaly, thick fibrinonecrotic membranes in the lumen of the large intestine, segmental darkening of the serosal surfaces of the small and large intestine, oronasal fistulae, and tongue and oral mucosa ulceration. On histology there is a heterophilic and lymphoplasmacytic infiltration of intestinal enteritis and hepatic necrosis. The inclusions are basophilic or amphophilic.[78,79] Adenovirus hepatic lesions have also been identified in individual cases (ornate box turtle (*Terrapene ornata ornata*).[76] Enlarged livers with hepatocellular vacuolar degeneration and rare intranuclear inclusions are the common findings.

Herpesvirus-associated infections has been described in a number of tortoises (California desert tortoise, spur-thigh tortoises, green sea turtles, Horsfield tortoises [*Testudo horsfieldii*], pancake tortoises [*Malacochersus tornieri*], Argentine tortoises [Geochelone chilensis], Mediterranean land tortoises, Hermann's and four-toed tortoises, Greek tortoises [*Testudo graeca*]).[80-82] This herpes virus infection is frequently complicated by secondary bacterial infections and generally results in severe debilitation of the tortoise. The clinical signs include nasal and ocular discharge, regurgitation, anorexia, and lethargy.[80] Necrotizing lesions of the oral cavity can be extensive including the tongue. The oral cavity, tongue, and into the esophagus may be covered with a yellowish-white pseudomembrane. It can progress to a tracheitis and bronchopneumonia and hepatitis. The liver can appear enlarged and supporting ecchymotic foci. Other organs involved supporting necrosis and intranuclear eosinophilic to amphophilic inclusion bodies can include spleen, stomach, cerebrum, lungs, adrenal glands, kidneys, duodenum, jejunum, colon, and pancreas. The transmission is most likely from infected carriers although the pathogenesis of this viral infection has not been fully described.

Infectious disease: bacterial

Any number of bacteria can be associated with a hepatitis. One unusual lesion was of a mycobacterial cholecystitis in a Pacific gopher snake (*Pituophis catenifer*).[83] The gall bladder was successfully removed and *Mycobacterium chelonae* isolated. The snake was treated with azithromycin for 6 months.

Infectious disease: parasitic

Whiptail lizards (*Cnemidophorus* spp.) from Texas and Colorado (USA), were examined for *Mesocestoides* sp. tetrathyridia and eleven (5%) were infected. Free tetrathyridia were found in the body cavity of lizards and encapsulated tetrathyridia were observed in the heart, liver, and stomach.[84]

Myxosporidia (protozoa of the class Myxosporea and phylum Myxozoa) are common parasites of fish, and uncommon parasites of reptiles. The genus *Myxidium* is best described in the urinary system of aquatic turtles (see previous discussion); however, these microbes are also recognized in the lumen of the gall bladder. Specific lesions due to the presence in the biliary system has not been described.[54] Definitive identification of these protozoa requires electron microscopy and genomic analysis.

Tumors

Metastases occur late in reptilian species. There is one report of a metastatic intestinal adenocarcinoma to the liver in an emerald tree boa (*Corallus canius*)[85] and the authors have identified metastases from gastric carcinomas into the intestine and liver of a carpet python (*Morelia spilota*) and in the liver of 2 bearded dragons (*Pogona vitticeps*).

There are several reports of a highly malignant gastric neuroendocrine carcinomas in young bearded dragons (*Pogona vitticeps*).[86,87] These tumors readily metastasize to the liver. The common clinical signs are anorexia, vomiting, hyperglycemia, and anemia.

The thyroid carcinoma metastasized to the liver and lungs in the rough knob-tail gecko (*Nephrurus amyae*).[88]

A C cell carcinoma of the thyroid gland in a crocodile lizard (*Shinisauris crocodilurus*) metastasized to the liver and mesentery.[89]

Malignant lymphoma (lymphosarcoma) represents the majority of the documented cases in reptiles. These neoplasms can involve many organs. In many reptiles it may initially present as cutaneous or oral tissue masses that are found to be metastatic disease. Grossly, they are gray-white or cream masses that are soft to relatively firm. Necrosis and hemorrhage may be present.[59,90,91]

Pancreas

Degenerative processes

A physiologic change that may fall under degenerative processes is of the atrophy of the intestinal mucosa and pancreas observed in snakes and chelonians, particularly those that have long periods of time between feeding.[92,93] There is reduction in the mass of the intestinal mucosa and depletion of the zymogen granules of the exocrine pancreas.[4,93] In studies of boas, pythons, and gila monsters (*Heloderma suspectum*) re-feeding resulted in hypertrophy of the pancreatic acinar and ductal cells as well as lengthening of the intestinal villi.[92,94]

Infectious disease

A variety of disease agents, from viruses to metazoans, have been reviewed in reptile pancreatic lesions, usually part of a more systemic disease.[95]

Tumors

Neoplasia of the endocrine pancreas is rare in reptiles. The few cases have been listed in surveys of reptile collections.[59,60,89] A captive born Savannah monitor (*Varanus exanthematicus*) had a functional metatastic islet cell tumor with elevated serum insulin. The tumor foci were in the liver (firm, white masses) and kidneys. The pancreas supported multiple nodular masses.[96] A splenopancreatic ductal adenocarcinoma in a northern water snake (*Nerodia sipedon*) produced multiorgan metastasis (stomach, liver, intravascular in the heart).[97] Grossly the tumor was a firm multinodular mass effacing the splenopancreas.

Adrenal Gland

Primary lesions of the adrenal gland are not reported in the literature. Left follicular 360-degree torsion in a green iguana involved the closely associated left adrenal gland.[98] Surgical removal was successful.

References

1. Zwart P. Renal pathology in reptiles. *Vet Clin North Am Exot Anim Pract.* 2006;9(1):129-159.

2. Innis CJ, Hawes M, Stone M, Seymour R. What's your diagnosis? - case 2: anamnesis. *Bull Assoc Reptilian Amphib Vet.* 1999;9(3):50-52.

3. Ariel E, Ladds PW, Buenviaje GN. Concurrent gout and suspected hypovitaminosis A in crocodile hatchlings. *Aust Vet J.* 1997;75(4):247-249.

4. Cowan DF. Diseases of captive reptiles. *J Am Vet Med Assoc.* 1968;153(7):848-859.

5. Garner MM, Gyimesi ZS, Rasmussen JM, et al. An amyloid-like deposition disorder in komodo dragons. *Proc Annu Conf Assoc Reptilian Amphib Vet.* 2003;50-51.

6. Hernandez-Divers SJ. Endoscopic renal evaluation and biopsy of Chelonia. *Vet Rec.* 2004;154(3):73-80.

7. Johnson JD. Urogenital system. In: Girling SJ, Raiti P, eds. *BSAVA Manual of Reptiles.* Gloucester, England: BSAVA;2004:261-272.

8. Richman L, Montali R, Allen M, Oftedal O. Widespread metastatic soft tissue mineralization in vitamin D deficient green iguana (*Iguana iguana*). *Vet Pathol.* 1995;32(5):560.

9. Shively MJ, Edwards K. Pathological calcification in a green iguana (*Iguana iguana*). *Avian Exot Pract.* 1985;2(1):32-33.

10. Cooper JE, Bloxam QMC, Tonge SJ. Pathology of round island geckos *Phelsuma geuntheri*: some unexpected findings. Dodo. *J Wildl Preserv Trusts.* 1998;34:153-158.

11. Garner MM, Lung NP, Murray S. Xanthomatosis in geckos: five cases. *J Zoo Wildl Med.* 1999;30(3): 443-447.

12. Montali RJ, Bush M, Smeller JM. The pathology of nephrotoxicity of gentamicin in snakes. A model for reptilian gout. *Vet Pathol.* 1979;16(1):108-115.

13. Casimire-Etzioni AL, Wellehan JFX, Embury JE, et al. Synovial fluid from an African spur-thighed tortoise (*Geochelone sulcata*). *Vet Clin Pathol.* 2004;33(1):43-46.

14. Miller HA. Urinary diseases of reptiles: Pathophysiology and diagnosis. *Semin Avian Exotic Pet Med.* 1998;7(2):93-103.

15. Wallach JD. Environmental and nutritional diseases of captive reptiles. *J Am Vet Med Assoc.* 1971;159(11):1632-1643.

16. Wallach JD, Hoessle C. Visceral gout in captive reptiles. *J Am Vet Med Assoc.* 1967;151(7):897-899.

17. Oftedal OT, Allen ME, Chung AL, et al. Nutrition, urates, and desert survival: potassium and the desert tortoise (*Gopherus agassizii*). *Proc Annu Conf Am Assoc Zoo Vet.* 1994;308-313.

18. Vancraeynest D, Pasmans F, Martel A, et al. Inclusion body disease in snakes: a review and description of three cases in boa constrictors in Belgium. *Vet Rec.* 2006;158(2):757-760.

19. Fleming GJ, Heard DJ, Jacobson ER, Buergelt C. Cytoplasmic Inclusions in Corn Snakes, Elaphe guttata, Resembling Inclusion Body Disease of Boid Snakes. *J Herpetol Med Surg.* 2003;13(2):18-22.

20. Raymond JT, Garner MM, Nordhausen W, Jacobson ER. A disease resembling inclusion body disease of boid snakes in captive palm vipers (*Bothriechis marchi*). *J Vet Diagn Invest.* 2001;13:82-86.

21. Schumacher J, Jacobson ER, Homer BL, Gaskin JM . Inclusion body disease in boid snakes. *J Zoo Wildlife Med.* 1994;25:511-524.

22. Carlisle-Nowak MS, Sullivan N, Carrigan M, et al. Inclusion body disease in two captive Australian pythons (*Morelia spilota variegata* and *Morelia spilota spilota*). *Aust Vet J.* 1998;76(2):98-100.

23. Stenglein MD, Jacobson ER, Chang L-W, et al. Widespread recombination, reassortment, and transmission of unbalanced compound viral genotypes in natural arenavirus infections. *PLoS Pathog.* 2015;11(5):e1004900.

24. Ariel E, Ladds PW, Roberts BL. Mycobacteriosis in young freshwater crocodiles (*Crocodylus johnstoni*). *Aust Vet J.* 1997;75(11):831-833.

25. Greer LL, Strandberg JD, Whitaker BR. *Mycobacterium chelonae* osteoarthritis in a Kemp's ridley sea turtle (*Lepidochelys kempii*). *J Wildl Dis.* 2003;39(3):736-741.

26. Suedmeyer WK, Gillespie DS, Pace L. Chromomycosis in a marine toad *Bufo marinus*. *Bull Assoc Reptilian Amphib Vet.* 1997;7(3):13-15.

27. Miller EA, Montali RJ, Ramsay EC, Rideout BA. Disseminated chromoblastomycosis in a colony of ornate-horned frogs (*Ceratophrys ornata*). *J Zoo Wildl Med.* 1992;23(4):433-438.

28. Tell L, Nichols DK, Bush M. Clinical challenge. *J Zoo Wildl Med.* 1994;25(1):173-175.

29. Jacobson ER. Chromomycosis and fibrosarcoma in a mangrove snake. *J Am Vet Med Assoc.* 1984;185(11): 1428-1430.

30. Joyner PH, Shreve AA, Spahr J, et al. Phaeohyphomycosis in a free-living eastern box turtle (*Terrapene carolina carolina*). *J Wildl Dis.* 2006;42(4):883-888.

31. Chiodini R, Sundberg JP. Styphlodora horrida in the kidneys and ureters of a boa constrictor (*Constrictor constrictor*). *Vet Med Small Anim Clin.* 1980;75(5):877-878.

32. Kazacos KR, Fisher LF. Renal styphlodoriasis in a boa constrictor. *J Am Vet Med Assoc.* 1977;171(9): 876-878.

33. Myers TS. Renal trematodiasis in a boa constrictor (*Constrictor constrictor*). *Lab Anim Sci.* 1971;21(3): 423-425.

34. Thatcher VE. A new species of Westella (Trematoda: Opisthogoniminae) from a Mexican snake. *J Parasitol.* 1963;49:123-124.

35. Innis CJ, Kincaid AL. Bilateral calcium phosphate ureteroliths and spirorchid trematode infection in a red-eared slider turtle, *Trachemys scripta elegans*, with a review of the pathology of spirorchiasis. *Bull Assoc Reptilian Amphib Vet.* 1999;9(3):32-35.

36. Johnson CA, Griffith JW, Tenorio P, Hytrek S, Lang CM. Fatal trematodiasis in research turtles. *Lab Anim Sci.* 1998;48(4):340-343.

37. Veazey RS, Stewart B, Snider TG. Ureteritis and nephritis in a Burmese python (*Python molurus bivitattus*) due to *Strongyloides* sp. infection. *J Zoo Wildl Med.* 1994;25(1);119-122.

38. Garner MM, Gardiner CH, Wellehan JF, et al. Intranuclear coccidiosis in tortoises: nine cases. *Vet Pathol.* 2006;43(3):311-320.

39. Jacobson, ER, Schumacher J, Telford SR, et al. Intranuclear coccidiosis in radiated tortoises (*Geochelone radiata*). *J Zoo Wildl Med.* 1994;25(1):95-102.

40. Reavill DR, Okimoto B, Barr BC, et al. Proliferative pneumonia due to intracellular protozoa in radiated tortoises (*Geochelone radiata*). *Proc Annu Conf Am Assoc Zoo Vet.* 2004.

41. Zwart P. Intraepithelial protozoon, *Klossiella boae* n. sp. in the kidneys of a boa constrictor. *J Protozool.* 1964;11:261-263.

42. Donaldson M, Heyneman D, Dempster R, Garcia L. 1975. Epizootic of fatal amebiasis among exhibited snakes: epidemiologic, pathologic, and chemotherapeutic considerations. *Am J Vet Res.* 36(6):807-817.

43. Jakob W, Wesemeier HH. Intestinal inflammation associated with flagellates in snakes. *J Comp Pathol.* 1995;112(4):417-421.

44. Kaneene JB, Taylor RF, Sikarskie JG, et al. Disease patterns in the Detroit Zoo: a study of reptilian and amphibian populations from 1973 through 1983. *J Am Vet Med Assoc.* 1985;87(11):1132-1133.

45. Kojimoto A, Uchida K, Horii Y, et al. Amebiasis in four ball pythons, *Python regius*. *J Vet Med Sci.* 2001;63(12):1365-1368.

46. MacNeill AL, Uhl EW, Kolenda-Roberts H, Jacobson E. Mortality in a wood turtle (*Clemmys insculpta*) collection. *Vet Clin Pathol.* 2002;31(3):133-136.

47. Hollamby S, Murphy D, Schiller CA. An epizootic of amoebiasis in a mixed species collection of juvenile tortoises. *J Herpetol Med Surg.* 2000;10(1): 9-15.

48. Jacobson E, Clubb S, Greiner E. Amebiasis in red-footed tortoises. J Am Vet Med Assoc. 1983;183(11): 1192-1194.

49. Jacobson ER, Green DE, Undeen AH, et al. Systemic microsporidiosis in inland bearded dragons (*Pogona vitticeps*). *J Zoo Wildl Med.* 1998;29(3):315-323.

50. Zwart P, Truyens EH. Hexamitiasis in tortoises. *Vet Parasitol.* 1975;1:175-183.

51. Garner MM, Bartholomew JL, Whipps CM, et al. Renal myxozoanosis in crowned river turtles *Hardella thurjii*: description of the putative agent *Myxidium hardella* n. sp. by histopathology, electron microscopy, and DNA sequencing. *Vet Pathol.* 2005;42(5):589-595.

52. Helke KL, Poynton SL. *Myxidium mackiei* (Myxosporea) in Indo-Gangetic flap-shelled turtles *Lissemys punctata andersonii*: parasite-host interaction and ultrastructure. *Dis Aquat Organ.* 2005;63(2-3):215-230.

53. Armed Forces Institute of Pathology. Wednesday Slide Conference. No. 28. Nichols D, ed. 1999. Available at: http://www.afip.org/vetpath/WSC/wsc98/98wsc28.htm.

54. Roberts JF, Whipps CM, Bartholomew JL, et al. *Myxidium scripta* n. sp. identified in urinary and biliary tract of Louisiana-farmed red-eared slider turtles *Trachemys scripta elegans*. *Dis Aquat Organ.* 2008;80(3): 199-209.

55. Barten SL, Davis K, Harris RK, Jacobson ER. Renal cell carcinoma with metastases in a corn snake (*Elaphe guttata*). *J Zoo Wildl Med.* 1994;25(1):123-127.

56. Burt DG, Gillett CS, Rush HG. Two cases of renal neoplasia in a colony of desert iguanas. *J Am Vet Med Assoc.* 1984;185(11):1423-1425.

57. Gravendyck M, Marschang RE, Schroder-Gravendyck AS, Kaleta EF. Renal adenocarcinoma in a reticulated python (*Python reticulatus*). *Vet Rec.* 1997;140(14):374-375.

58. Jacobson ER, Long PH, Miller RE, et al. Renal neoplasia of snakes. *J Am Vet Med Assoc.* 1986;189(9): 1134-1136.

59. Hernandez-Divers SM, Garner MM. Neoplasia of reptile with an emphasis on lizards. *Vet Clin North Am Exot Anim Pract.* 2003;6:251-273.

60. Garner MM, Hernandez-Divers SM, Raymond JT. Reptile neoplasia: a retrospective study of case submissions to a specialty diagnostic service. *Vet Clin North Am Exot Anim Pract.* 2004;7:653-671.

61. Kwantes LJ. Surgical correction of cystic urolithiasis in an iguana. *Can Vet J.* 1992;33:752-753.

62. Mader D, Ling GV, Ruby AL. Cystic calculi in the California desert tortoise: evaluation of 100 cases. *Proc 6th Int Symp Pathol Reptiles Amphib.* 2002;177-178.

63. McKown RD. A cystic calculus from a wild western spiny softshell turtle (*Apalone* [*Trionyx*] *spiniferus hartwegi*). *J Zoo Wildl Med.* 1998;29(3):347.

64. Lightfoot T. Bladder necrosis secondary to cystic calculus in a green iguana. *Exot DVM.* 1999;1(3):29-33.

65. Miller SM, Serrano MA, Garner MM. Ectopic ovum in a hermaphrodite common green iguana, *Iguana iguana*. *J Herpetol Med Surg.* 2003;13(4):23-26.

66. Sharpe S, Lamm CG, Killick R. Intracoelomic anaplastic sarcoma in an intersex Madagascar tree boa (*Sanzinia madagascariensis*). *J Vet Diagn Invest.* 2013;25(1):153-157.

67. Sepulveda MS, Wiebe JJ, Honeyfield DC, et al. Organochlorine pesticides and thiamine in eggs of large-mouth bass and American alligators and their relationship with early life-stage mortality. *J Wildl Dis.* 2004;40(4):782-786.

68. Pees M, Ludewig E, Plenz B, Schmidt V. Imaging diagnosis-seminoma causing liver compression in a spur-thighed tortoise (*Testudo graeca*). *Vet Radiol Ultrasound.* 2015;56(2):E21-E24.

69. Anderson NL, Williams J, Sagartz JE, Barnewall R. Ovarian teratoma in a green iguana (*Iguana iguana*). *J Zoo Wildl Med.* 1996;27(1):90-95.

70. Newman SJ, Brown CJ, Patnaik AK. Malignant ovarian teratoma in a red-eared slider (*Trachemys scripta elegans*). *J Vet Diagn Invest.* 2003;15(1):77-81.

71. Petterino C, Bedin M, Podest G, Ratto A. Undifferentiated tumor in the ovary of a corn snake (*Elaphe guttata guttata*). *Vet Clin Pathol.* 2006;35(1):95-100.

72. Roberts WG, Klein MK, Loomis M, et al. Photodynamic therapy of spontaneous cancers in felines, canines, and snakes with chloro-aluminum sulfonated phthalocyanine. *J Natl Cancer Inst.* 1991;83(1):18-23.

73. Palmieri C, Selleri P, Di Girolamo N, et al. Multiple congenital malformations in a dicephalic spur-thighed tortoise (*Testudo graeca ibera*). *J Comp Pathol.* 2013;149(2-3):368-371.

74. Andreadis PT, Burghardt GM. Feeding behavior and an oropharyngeal component of satiety in a two-headed snake. *Physiol Behav.* 1993;54(4):649-658.

75. Ball I, Ofner S, Funk RS, et al. Prevalence of neutralising antibodies against adenoviruses in lizards and snakes. *Vet J.* 2014;202(1):176-181.

76. Farkas SL, Harrach B, Benkö M. Completion of the genome analysis of snake adenovirus type 1, a representative of the reptilian lineage within the novel genus Atadenovirus. *Virus Res.* 2008;132(1-2):132-139.

77. Papp T, Fledelius B, Schmidt V, et al. PCR-sequence characterization of new adenoviruses found in reptiles and the first successful isolation of a lizard adenovirus. *Vet Microbiol.* 2009;134(3-4):233-240.

78. Doszpoly A, Wellehan Jr, JFX, Childress AL, et al. Partial characterization of a new adenovirus lineage discovered in testudinoid turtles. *Infect Genet Evol.* 2013;17:106-112.

79. Rivera S, Wellehan JF, McManamon R, et al. Systemic adenovirus infection in Sulawesi tortoises (*Indotestudo forsteni*) caused by a novel Siadenovirus. *J Vet Diagn Invest.* 2009;21:415-426.

80. Jacobson ER, Clubb S, Gaskin JM, Gardiner C. Herpesvirus-like infection in Argentine tortoises. *J Am Vet Med Assoc.* 1985;187(11):1227-1229.

81. Cooper JE, Gschmeissner S, Bone RD. Herpes-like virus particles in necrotic stomatitis of tortoises. *Vet Rec.* 1988;123(21):544.

82. Une Y, Uemura K, Nakano Y, et al. Herpesvirus infection in tortoises (*Malacochersus tornieri* and *Testudo horsfieldii*). *Vet Pathol.* 1999;36:624-627.

83. Kinney ME, Chinnadurai SK, Wack RF. Cholecystectomy for the treatment of mycobacterial cholecystitis in a Pacific gopher snake (*Pituophis catenifer*). *J Herpetol Med Surg.* 2013;23(1-2):10-14.

84. McAllister CT, Cordes JE, Conn DB, et al. Helminth parasites of unisexual and bisexual whiptail lizards (Teiidae) in North America. V. *Mesocestoides* sp. tetrathyridia (Cestoidea: Cyclophyllidea) from four species of *Cnemidophorus*. *J Wildl Dis.* 1991;27:494-497.

85. Oros J, Lorenzo H, Andrada M, Recuero J. Type A-like retroviral particles in a metastatic intestinal adenocarcinoma in an emerald tree boa (*Corallus caninus*). *Vet Pathol.* 2004;41:515-518.

86. Ritter JM, Garner MM, Chilton JA, et al. Gastric neuroendocrine carcinomas in bearded dragons (*Pogona vitticeps*) *Vet Pathol.* 2009;46:1109-1116.

87. Lyons JA, Newman SJ, Greenacre CB, Dunlap J. A gastric neuroendocrine carcinoma expressing somatostatin in a bearded dragon (*Pogona vitticeps*). *J Vet Diagn Invest.* 2010; 22:316-320.

88. Hadfield CA, Clayton LA, Clancy MM, et al. Proliferative thyroid lesions in three diplodactylid geckos: *Nephrurus amyae*, *Nephrurus levis*, and *Oedura marmorata*. *J Zoo Wildl Med.* 2012;43(1):131-140.

89. Sykes JM, Trupkiewicz JG. Reptile neoplasia at the Philadelphia Zoological Garden, 1901-2002. *J Zoo Wildl Med.* 2006;37(1):11-19.

90. Effron M, Griner L, Benirschke K. Nature and rate of neoplasia found in captive wild mammals, birds, and reptiles at necropsy. *J Natl Cancer Inst.* 1977;59(1):185-198.

91. Oros J, Torrent A, Espinosa De Los Monteros A, et al. Multicentric lymphoblastic lymphoma in a loggerhead sea turtle (*Caretta caretta*). *Vet Pathol.* 2001;38:464-467.

92. Cox CL, Secor SM. Matched regulation of gastrointestinal performance in the Burmese python, *Python molurus*. *J Exp Biol.* 2008;211:1131-1140.

93. Zhalka M, Bdolah A. Dietary regulation of digestive enzyme levels in the water snake, *Natrix tessellata*. *J Exp Zool.* 1987;243(1):9-13.

94. Christel CM, DeNardo DF, Secor SM. Metabolic and digestive response to food ingestion in a binge-feeding lizard, the gila monster (*Heloderma suspectum*). *J Exp Biol*. 2007; 210:3430-3439.

95. Stahl SJ. Diseases of the reptile pancreas. *Vet Clin North Am Exot Anim Pract.* 2003;6(1):191-212.

96. Naples LM, Langan JN, Mylniczenko ND, et al. Islet cell tumor in a Savannah monitor (*Varanus exanthematicus*). *J Herpetol Med Surg.* 2009;19(4):97-105.

97. Matthews ME, Eshar D, Delk KW, et al. Splenopancreatic ductal adenocarcinoma with multiorgan metastasis in a northern water snake (*Nerodia sipedon*). *J Zoo Wildl Med.* 2014;45(2):437-440.

98. Mehler SJ, Rosenstein DS, Patterson JS. Imaging diagnosis-follicular torsion in a green iguana (*Iguana iguana*) with involvement of the left adrenal gland. *Vet Radiol Ultrasound.* 2002;43(4):343-345.

Diagnostics and Treatment of the Caudal Coelom

J. Jill Heatley, DVM, MS, DABVP (Avian Practice), DABVP (Reptile and Amphibian Practice), Dipl ACZM, and
Cameron Ratliff, DVM

Session #2039

Affiliation: From Texas A&M University, College of Veterinary Medicine and Biomedical Sciences, University Dr, College Station, TX 77843, USA.

Abstract: Diagnostics and treatment to include expected history & clinical signs of affected reptiles, clinical pathology, imaging and medical therapy of the common diseases and dysfunction of the organs of the caudal coelom will be reviewed based on author clinical practical knowledge and literature review. Organs reviewed will include liver/gall bladder, pancreas, intestines, cloaca, kidneys/urinary tract, reproductive organs and the coelom itself.

Liver and Gall Bladder

Clinical signs of liver and /or gall bladder disease are often nonspecific. Reported findings include those related to appetite including a gradual reduction, acute onset or post hibernation anorexia as well as less commonly diarrhea or regurgitation may occur. Other general clinical signs include reduced activity, apathy, sudden onset depression, lethargy and weakness.[1] Reduced fecundity and fertility, hibernation problems. Gradual weight loss, retarded weight gain, loss of body condition or changes in body mass are commonly reported.[2-4] Changes in color of feces, mucous membranes or other tissues may occur including jaundice, yellow green urates and pale, or hyperemic mucous membranes.[2,3] Plasma, lymph, and tissues may be discolored yellow or green in amphibians in which biliverdin and bilirubin both occur from heme breakdown. Thus increased plasma bilirubin may occur in cases of cholangiohepatitis and chronic fibrosing hepatitis of some amphibians.[4]

For determination of liver size and density, few reference intervals have been established.[4] CT and MRI may be more valuable than radiographs based on poor coelomic contrast in most reptiles.[1] Endoscopic or ultrasound guided percutaneous aspiration or biopsy techniques have been described for most reptiles.[5-7]

Lack of baseline data in reptiles and amphibians as well as enzyme variability between species in reptiles limits the usefulness of clinical pathologic diagnosis of liver disease in these species. Abnormal clinicopathologic values from a recent case of a tortoise, *Testudo graeca* with a seminoma compressing the liver demonstrates analytes that may be useful for diagnosis of liver disease in reptiles: decreased serum protein levels (total proteins 19.0 g/L; albumin 7.8 g/L) and massively increased enzyme levels (aspartate transaminase (AST) 5103 U/L; glutamate dehydrogenase (GLDH) 270 U/L; lactate dehydrogenase (LDH) 2866 U/L.[1] Despite severe hepatic infiltration of fat and tumor of the liver in a bearded dragon (*Pogona vitticeps*), AST (25 U/L), bile acids (58 µmol/L), cholesterol (284 mg/dl), and triglycerides (43 mg/dl) were all considered within reference intervals.[8] Repeated carprofen treatment of iguanas (*Iguana iguana*) resulted in increased concentrations of ALT.[9] Few studies characterize enzymes found in the liver of reptile and amphibian species and these are often not confined to the liver and vary across species. AST, LDH, and ALT most commonly occur in diagnostically acceptable amounts while GGT and GMD are generally not of diagnostic value for liver disease.[10-15] Bilirubin values may be of use for amphibians;[4] however biliverdin, the most common heme break down product of reptiles, is not a commercially available assay and has a very short half-life.

Bile acids (BA) have been scientifically validated in relatively few reptile and amphibian species. Nonetheless they are commonly justified as diagnostic of hepatic disease should values be > 60 μmol/L.[2] Bile acids may increase based on dietary intake, so increased values from the fasted or anorexic reptile may be of more diagnostic import. Mean plasma BA concentration in plasma samples from healthy green iguanas fasting 24 hours was 15.89 ± 15.61 μmol/L and fasting for 48 hours was 9.56 ± 8.52 μmol/L.[16] Bile acids (3α-hydroxy BA) increase in plasma of green iguanas (*I. iguana*) with chronic liver diseases with the highest concentrations (> 70 μmol/L) present in patients suffering from liver cirrhosis, liver lipidosis and liver neoplasms.[17] The validity, sensitivity and specificity of bile acids as a diagnostic for liver disease in all reptiles remains suspect as demonstrated by a recent case report of hepatic encephalopathy in a red-tailed boa (*Boa constrictor imperator*) with a plasma bile acids of < 35 μmol/L.[18]

Treatment of liver disease is generally supportive care: preferred optimum temperature zone, fluid therapy and nutritional support should be provided while, antimicrobials, antiinflammatory drugs, and allopurinol should be considered. In cases of obesity and hepatic lipidosis, carnitine, choline, methionine, thyroxine and anabolic steroids may be considered to facilitate fat metabolism and improve intake. Nutraceuticals used for hepatic lipidosis in an Atlantic Ridley sea turtle (*Lepidocheyls kempii*) included artichoke and milk thistle, ALA, lecithin, and omega-3 fatty acids resulted in improved in serum biochemistry liver values and behaviors.[19]

Pancreas

Documented specific clinical signs in reptiles attributable to pancreatic disease are few as the disease appears relatively rare. In red eared sliders, *Trachemys scripta elegans*, parasitic migration in the pancreas causes necrosis, leukocytic infiltrate and destruction of exocrine tissue. The blood profile of 1 of 3 turtles revealed eosinophilia and hyperglycemia;[20] however, hyperglycemia of reptiles has been variably associated with pancreatic disease but may also be a normal physiologic response in many species based on arousal from hibernation, prolonged inanition combined with post prandial changes, as a cryoprotectant and based on stress, or neoplasia (gastric neuroendocrine carcinoma).[21] Thus increased blood glucose is not specific for pancreatic disease. Amylase and lipase have been evaluated in a number of reptiles. Further, in Dumeril's monitors (*Varanus dumerili*) amylase population-based ranges should be assessed with caution based on individual animal variation.[22] Ultrasound diagnosis of pancreatitis also remains undocumented in reptiles. Similarly treatment options are limited as attempted treatment of hyperglycemia with insulin or other hypoglycemic agents in reptiles appears to have minimal effects.[21]

Gastrointestinal Tract: Intestines

Intestinal diseases reported in reptiles include strictures (sea turtles), enteritis, foreign body ingestion, parasitism and cryptosporidiosis. Bacterial enteritis is likely common in reptiles; however, diagnosis may be hindered by lack of knowledge of normal flora of the species and species bacterial flora natural variety.[23-26] Anaerobic bacterial infection, unculturables and viral infection should also be considered.[27] Cryptosporidial infection occurs in reptiles, diagnosis may be challenging and is based on acid fast staining or PCR. The amount of acceptable or expected gas in the reptilian GI tract has not been defined and likely varies based on the diet. Thus plain radiographs are seldom diagnostic for enteritis as a single entity in reptiles. Contrast radiography be more useful for assessment of GI transit, and foreign body ingestion / obstruction/ obstipation.[28] Unlike in companion mammal medicine clinical pathologic tests to evaluate chronic intestinal inflammation are not yet applicable for reptiles.

Gastrointestinal Tract: Cloaca

Cloaca, cloacal prolapse

History and clinical signs for cloacal prolapse are self-evident, however determination of the time of prolapse is key in facilitating prognosis and determining if medical management remains a viable option.

Diseases affecting the cloaca, as a complex structure involving the urogenital and gastrointestinal tracts are numerous. In addition, systemic or adjacent diseases may contribute to cloacal disease. Cloacal prolapse is not an uncommon presentation in private practice and has been reported at a rate of ~2% of reptile cases.[29] Technically cloacal prolapse is a misnomer or catch-all term as something is generally prolapsed from the cloaca. Colon prolapses appear more likely to occur in lizards, bladder prolapses more commonly in tortoises and cloacal prolapses appear rare in snakes.[29] Recurrent prolapses are most often associated with reproductive or metabolic disease.[29] Identification of prolapsed tissue is key in appropriate treatment of these cases, which may be successfully managed medically or surgically. Radiographs are indicated as well as fecal examination and CBC and plasma chemistries to presurgically assess the patient and to rule out common causes of cloacal prolapse: urinary calculi, metabolic bone disease (nutritional or renal), constipation, obstipation, reproductive disease, eggs, and parasitism. Cryptosporidiosis has also been associated with cloacitis and prolapse in the iguana.[30]

Urinary System

Renourinary diagnostic analytes commonly assessed in reptiles include uric acid, and urea (more commonly assessed in tortoises). Both of the analytes are increased based on dehydration, and uric acid may be increased postprandially.[31] Should increases in these analytes remain when normal hydration is established, renal disease should be considered. Electrolytes (potassium increase) and proteins may also be useful but may be more difficult to interpret based on lack of reference intervals and findings in reptiles suffering renal disease. In females, plasma concentrations of proteins, calcium, and phosphorous may increase based on reproductive activity but may also change with renal disease and diet. Packed cell volume, Hb, RBC may be useful to assess dehydration and may be decreased in cases of chronic renal disease. Urinalysis has received little attention in reptiles, likely based on the lack of bladder sterility and the inability to diagnose kidney disease based on urine specific gravity; however urine sediment and metabolites still have promise as viable diagnostics in these species.[32] Normal values for some species have been reported.

Radiographs may be useful for the diagnosis of urinary tract disease. Kidney enlargement and urolithiasis (which may confirm palpation suspicions), bone density, soft tissue mineralization, constipation (renomegalic impingement) may be visualized. CT may facilitate diagnosis particularly within chelonia. Most renal diagnostic references concentrate on diseases of iguanas which are becoming less prevalent in the pet trade.[33,34] Intravenous urography (Conray 280, others) can be very useful when attempting to identify renal changes (abscesses, neoplasia, calculi), renal and ureteral damage and ureteral obstructions.[35] Treatment of renal disease is based on establishment of normal hydration, electrolyte, mineral and blood protein values. Calcium glubionate, gluconate, calcium carbonate, vitamin D^3, calcitonin and dietary modification may be used in animals with hyperphosphatemia and hypercalcemia.[35] Diuretics include furosemide and hydrochlorothiazide. Allopurinol can reduce uric acid production. Steroids administration may decrease nephrocalcinosis (prednisolone) and decrease protein catabolism (nandrolone).

Reproductive System

Most clinical signs of reproductive disease are related to female diseases and the limited capacity of the cloacal orifice. Prolapse of reproductive organs may be the most obvious clinical sign, however coelomic swelling, anorexia, and lethargy or hyperactivity based on the need to lay may also occur. Coelomic cytology may elucidate serious disease such as rupture of the reproductive tract. Increased concentrations of blood calcium, phosphorous, albumin, globulins and triglycerides may indicate reproductive activity, but do not rule disease in or out. Radiographs may also be useful; however in the case of reproductive neoplasia, ultrasound or CT may provide a more complete presurgical working diagnosis.[36] Treatment is generally supportive care and removal (or repair and replacement) of the offending organ. Note that in the case of hemipene or phallus devitalization, organ resection is acceptable and will not interfere with intromission (hemipenes) or urate production.

Coelomitis

Causes of coelomitis reported in the literature include yolk coelomitis, urolith ceolomitis, renal nephrolithiasis with secondary ductular rupture, and renal infection and idiopathic etiology/sepsis.[32,37-40] Coelomitis would also be expected secondary to or in conjunction with sepsis and compromise of the gastrointestinal tract and presumed bacterial leakage or transmural location, or compromise to the renal system, particularly the bladder which of course is not sterile.[41-44] The most common infectious culprits are bacterial *Salmonella* spp., *Citrobacter* spp., or renal infection with flagellated protozoa.[32,41, 45-47] History and clinical signs referable to coelomitis are vague and few although signs associated with sepsis (petechial) shock or swelling of the coelom and fluid based on coelomic tap or ballotment may be present. Fluid lines upon radiograph or free fluid based on ultrasound may suggest coelomitis, especially if it is flocculent;[48] however, bladder enlargement and urates in the bladder should be ruled out. Expected clinical pathology findings specific to coelomitis are limited to cytologic examination and fluid analysis of coelomic aspirates; however, clinical pathology findings consistent with sepsis could also be indicative of coelomitis. Specific treatment of ceolomitis is poorly documented for reptiles. Surgical exploration and removal of offending structures and lavage or systemic treatment with broad spectrum antibiotics and anti-inflammatories should be considered.

References

1. Pees M, Ludewig E, Plenz B, Schmidt V. Imaging diagnosis–seminoma causing liver compression in a spur-thighed tortoise (*Testudo graeca*). *Vet Radiol Ultrasound.* 2015;56(2):E21-E24.

2. Divers SJ. Reptilian liver and gastrointestinal testing. In: Fudge AM, ed. *Laboratory Medicine: Avian and Exotic Pets.* Philadelphia, PA: WB Saunders;2000:205-209.

3. Divers SJ, Cooper JE. Reptile hepatic lipidosis. *Semin Avian Exot Pet Med.* 2000;9(3):153-164.

4. Crawshaw GJ, Weinkle TK. Clinical and pathological aspects of the amphibian liver. *Semin Avian Exot Pet Med.* 2000;9(3):165-173.

5. Isaza R, Ackerman N, Schumacher J. Ultrasound-guided percutaneous liver biopsy in snakes. *Vet Radiol Ultrasound.* 1993;34(6):452-454.

6. Isaza R, Ackerman N, Jacobson ER. Constrictor (*Boa constrictor*). *Vet Radiol Ultrasound.* 1993;34(6): 445-50.

7. Sainsbury AW, Gili C. Ultrasonographic anatomy and scanning technique of the coelomic organs of the Bosc monitor (*Varanus exanthematicus*). *J Zoo Wildl Med.* 1991;421-33.

8. Perpiñán D, Addante K, Driskell E. Gastrointestinal disturbances in a bearded dragon (*Pogona vitticeps*). *J Herpetol Med Surg.* 2010;20(2):54-57.

9. Trnkova S, Knotkova Z, Hrda A, Knotek Z. Effect of non-steroidal anti-inflammatory drugs on the blood profile in the green iguana (*Iguana iguana*). *Veterinarni Medicina-praha.* 2007;52(11):507.

10. Ramsay EC, Dotson TK. Tissue and serum enzyme activities in the yellow rat snake (*Elaphe obsoleta quadrivitatta*). *Am J Vet Res.* 1995;56(4):423-428.

11. Wagner RA, Wetzel R. Tissue and plasma enzyme activities in juvenile green iguanas. *Am J Vet Res.* 1999;60(2):201-203.

12. Anderson ET, Socha VL, Gardner J, et al. Tissue enzyme activities in the loggerhead sea turtle (*Caretta caretta*). *J Zoo Wildl Med.* 2013;44(1):62-69.

13. Petrosky KY, Knoll JS, Innis C. Tissue enzyme activities in Kemp's ridley turtles (*Lepidochelys kempii*). *J Zoo Wildl Med.* 2015;46(3):637-640.

14. Bogan JE, Mitchell MA. Characterizing tissue enzyme activities of Cuban tree frogs (*Osteopilus septentrionalis*). *J Herpetol Med Surg.* 2017. In press.

15. Bogan Jr JE, Mitchell MA. Characterizing tissue enzyme activities in the American alligator (*Alligator mississippiensis*). *J Herpetol Med Surg.* 2014;24(3):77-81.

16. McBride M, Hernandez-Divers SJ, Koch T, et al. Preliminary evaluation of pre-and post-prandial 3α-hydroxy-bile-acids in the green iguana (*Iguana iguana*). *J Herpetol Med Surg.* 2007;16:129-134.

17. Knotek Z, Knotkova Z, Hrda A, Dorrestein GM. Plasma bile acids in reptiles. *Proc Annu Conf Assoc Reptilian Amphib Vet.* 2009;124-127.

18. Di Giuseppe M, Oliveri M, Morici M, et al. Hepatic encephalopathy in a red-tailed boa (*Boa constrictor imperator*). *J Exot Pet Med.* 2017;26(2):96-100.

19. Ahlstrom RT, Wolf T, Peterson D, Pestano N, Mejia-Fava J. Alternative treatment options for managing hepatic lipidosis in an Atlantic Ridley sea turtle (*Lepidocheyls kempii*). *Proc Annu Conf Int Assoc Aquat Anim Med.* 2012.

20. Hidalgo-Vila J, Martínez-Silvestre A, Ribas A, et al. Pancreatitis associated with the helminth *Serpinema microcephalus* (Nematoda: Camallanidae) in exotic red-eared slider turtles (*Trachemys scripta elegans*). *J Wildl Dis.* 2011;47(1)201-205.

21. Stahl SJ. Hyperglycemia in reptiles. In: Mader D, ed. *Reptile Medicine and Surgery.* St Louis, MO: Elsevier Saunders;2006:822-830.

22. Bertelsen MF, Kjelgaard-Hansen M, Howell JR, Crawshaw GJ. Short-term biological variation of clinical chemical values in Dumeril's monitors (*Varanus dumerili*). *J Zoo Wildl Med.* 2007;38(2):217-221.

23. Mitchell MA, Diaz-Figueroa O. Clinical reptile gastroenterology. *Vet Clin North Am Exot Anim Pract.* 2005;8(2):277-298.

24. Seixas R, Pissarra H, Santos J, et al. Severe fibrinonecrotic enteritis caused by *Pseudomonas aeruginosa* in a captive monitor lizard (*Varanus niloticus*). *J Zoo Wildl Med.* 2014;45(2):410-412.

25. Martínez-Silvestre A, Guinea D, Ferrer D, Pantchev N. Parasitic enteritis associated with the camallanid nematode *Serpinema microcephalus* in wild invasive turtles (*Trachemys*, *Pseudemys*, *Graptemys*, and *Ocadia*) in Spain. *J Herpetol Med Surg.* 2015;25(1-2):48-52.

26. Romero SB, Čížek A, Masaříková M, Knotek Z. Choanal and cloacal aerobic bacterial flora in captive green iguanas: a comparative analysis. *Acta Veterinaria Brno.* 2015;19;84(1):19-24.

27. Szirovicza L, López P, Kopena R,et al. Random sampling of squamate reptiles in Spanish natural reserves reveals the presence of novel adenoviruses in lacertids (family Lacertidae) and worm lizards (Amphisbaenia). *PLoS One.* 2016;11(7):e0159016.

28. Mans C. Clinical update on diagnosis and management of disorders of the digestive system of reptiles. *J Exot Pet Med.* 2013;22(2):141-162.

29. Hedley J, Eatwell K. Cloacal prolapses in reptiles: a retrospective study of 56 cases. *J Small Anim Pract.* 2014;55(5):265-268.

30. Kik MJ, van Asten AJ, Lenstra JA, Kirpensteijn J. Cloaca prolapse and cystitis in green iguana (*Iguana iguana*) caused by a novel *Cryptosporidium* species. *Vet Parasitol.* 2011;175(1):165-167.

31. Maixner JM, Ramsay EC, Arp LH. Effects of feeding on serum uric acid in captive reptiles. *J Zoo Anim Med.* 1987;18(2-3):62-65.

32. Juan-Sallés C, Garner MM, Nordhausen RW, et al. Renal flagellate infections in reptiles: 29 cases. *J Zoo Wildl Med.* 2014;45(1):100-109.

33. Knotek Z, Hauptman K, Knotkova Z, et al. Renal disease haemogram and plasma biochemistry in green iguana. *Acta Veterinaria Brno.* 2002;71(3):333-340.

34. Knotek Z, Dorrestein GM, Knotková Z, Zwart P. Chronic renal failure disease in adult green iguanas (*Iguana iguana*). *Proc Annu Conf Br Vet Zool Soc.* 2009;7(8):11.

35. Hernandez-Divers SJ, Stahl SJ, Stedman NL, et al. Renal evaluation in the healthy green iguana (*Iguana iguana*): assessment of plasma biochemistry, glomerular filtration rate, and endoscopic biopsy. *J Zoo Wildl Med.* 2005;36(2):155-168.

36. Gumpenberger M. Diagnostic imaging of reproductive tract disorders in reptiles. *Vet Clin North Am Exot Anim Pract.* 2017;20(2):327-343.

37. Stacy BA, Howard L, Kinkaid J, et al. Yolk coelomitis in Fiji Island banded iguanas (*Brachylophus fasciatus*). *J Zoo Wildl Med.* 2008;39(2):161-169.

38. Goe A, Heard DJ, Fredholm DV, et al. Management of a patent urachus and yolk coelomitis in a prehensile-tailed skink (*Corucia zebrata*). *J Zoo Wildl Med.* 2015;46(4):909-912.

39. Gardner BR, Barrows MG. Yolk coelomitis in a white-throated monitor lizard (*Varanus albigularis*). *J South Afr Vet Assoc.* 2010;81(2):121-122.

40. Lair S, Couture ÉL, Ferrell ST, et al. Gastro-duodenal perforations and coelomitis in a group of juvenile spiny softshell turtles (*Apalone spinifera*). *J Herpetol Med Surg.* 2017. In press.

41. Goldberg DW, Cegoni CT, Rogério DW, et al. Fatal *Citrobacter* coelomitis in a juvenile green turtle (*Chelonia mydas*): a case report. *Mar Turtle Newsl.* 2016;150:10.

42. Ebani VV, Fratini F. Bacterial zoonoses among domestic reptiles. *Annali della Facoltà di Medicina Veterinaria.* 2005;58:85-91.

43. Jankowski G, Sirninger J, Borne J, Nevarez JG. Chemotherapeutic treatment for leukemia in a bearded dragon *(Pogona vitticeps)*. *J Zoo Wildl Med.* 2011;42(2):322-325.

44. Oros J, Rodriguez JL, Herráez P, et al. Respiratory and digestive lesions caused by *Salmonella arizonae* in two snakes. *J Comp Pathol.* 1996;115(2):185-189.

45. Dunbar JP, Zarelli M, Martin SA, et al. Trunk vertebrae osteomyelitis in a spectacled caiman (*Caiman crocodilus*). *Herpetol Bull.* 2015;134.

46. Gartrell BD, Youl JM, King CM,et al. Failure to detect *Salmonella* species in a population of wild tuatara (*Sphenodon punctatus*). *NZ Vet J.* 2007;55(3):134-136.

47. Le Souëf AT, Barry M, Brunton DH, et al. Ovariectomy as treatment for ovarian bacterial granulomas in a Duvaucel's gecko (*Hoplodactylus duvaucelii*). *NZ Vet J.* 2015;63(6):340-344.

48. Music MK, Strunk A. Reptile critical care and common emergencies. *Vet Clin North Am Exot Anim Pract.* 2016;19(2):591-612.

Endoscopy of the Caudal Coelom: Gastrointestinal, Hepatobiliary, Pancreatic, and Urinary Systems

Stephen J. Divers, BVetMed, DZooMed, DipECZM (Herpetology), DipECZM (Zoo Health Management), Dipl ACZM, FRCVS

Session #2100

Affiliation: From Department of Small Animal Medicine and Surgery (zoological medicine), College of Veterinary Medicine, University of Georgia, Athens, GA 30602, USA.

Introduction

A definitive diagnosis is important to maximize treatment success, but reptile cases are infrequently mismanaged simply because of a failure to identify a specific problem. Definitive diagnosis relies upon the demonstration of a host pathologic response (by histopathology, less reliably cytology, or using paired rising antibody titers), and the causative agent (by microbiology, parasitology, or toxicology). There are few reliable serologic tests available for reptiles, and those that are available require a minimum of 6-9 weeks to demonstrate the necessary two-fold increase in titers. It is therefore clear that tissue samples offer the most expedient means to a diagnosis, and endoscopy offers a visually-accurate, minimally-invasive, ante-mortem method to collect such material. Recent reviews on the topic of reptile endoscopy and endosurgery have been published.[1,2]

Rigid Endoscopy Equipment and Use

Despite the advances in flexible endoscopy equipment, rigid endoscopy still maintains its central role in reptile diagnostics largely because of its abilities to gain access to most areas of interest in most reptiles; however, given the variation in reptile size, species-specific anatomy, and variety of procedures that may be performed, a selection of different endoscopes and instruments is recommended, which can then be tailored to suit individual practice needs (Table 1).

Saline and CO_2 insufflation

Some form of insufflation is required for most coelioscopic procedures. CO_2 gas delivered by a dedicated endoflator is preferred, but air delivered by syringe or a small aquarium air pump has been used with limited success. The risks associated with the use of air, specifically air emboli, must not be ignored but appear rare. Dedicated endoflators precisely maintain the set insufflation pressure within the coelom by matching gas in-flow with any leakage. In some situations, the use of sterile saline can be especially helpful for creating mild distension. Saline infusion is particularly useful when dealing with a hollow viscus (eg, gastrointestinal tract, bladder, cloaca), or when performing coelioscopy in small reptiles (eg, neonate chelonian gender identification). Saline infusion may also be preferred for coelioscopy of certain aquatic reptiles where there are potential ill-effects of residual insufflation gas on post-operative buoyancy.

Table 1. Endoscopic instrumentation for reptiles.

Equipment description	Primary indications
Visualization and documentation	
Endovideo camera and monitor Xenon light source and light guide cable Digital capture device (eg, AIDA-DVD)	Required for all endoscopy procedures
Rigid telescopes and endoscopes	
1 mm × 20 cm semi-rigid miniscope, 0°	Stomatoscopy, otoscopy, rhinoscopy, tracheoscopy in animals up to 500 g
1.9 mm × 18.5 cm telescope, 30° oblique, with integrated 3.3 mm operating sheath	Stomatoscopy, otoscopy, rhinoscopy, tracheoscopy, gastroscopy, colonoscopy, cloacoscopy, and coelioscopy in animals up to 2-3 kg
2.7 mm × 18 cm telescope, 30° oblique 4.8 mm operating sheath	Stomatoscopy, otoscopy, rhinoscopy, tracheoscopy, gastroscopy, colonoscopy, cloacoscopy, and coelioscopy in animals between 100 g and 20 kg
5 mm × 29 cm telescope, 0°, with 6 mm × 6.5 cm Ternamian EndoTip cannula	Coelioscopy of animals over 20 kg
5 mm × 29 cm telescope, 30° oblique, with 6 mm × 6.5 cm Ternamian EndoTip cannula	Coelioscopy of animals, especially chelonians, over 20 kg
5 mm × 8.5 cm otoendoscope, 0°, with integrated operating sheath	Stomatoscopy and otoscopy in animals between 1 and 50 kg
Flexible instruments for use with rigid telescopes and operating sheaths	
1-mm biopsy forceps 1-mm grasping forceps	For use with 1.9-mm telescope and integrated sheath
1.7-mm biopsy forceps 1.7-mm single-action scissors 1.7-mm remote injection needle 1.7-mm grasping/retrieval forceps 1.7-mm wire basket retrieval 1.7-mm needle end radiosurgery electrode 1.7-mm polypectomy snare	For use with 2.7-mm telescope and 4.8 mm operating sheath
Insufflation	
Dedicated CO_2 insufflator, capable of insufflation pressures as low as 3 mmHg and flow rates as low as 0.1 L/min; CO_2 compressed gas with regulator; silicone supply tubing to endoscopy equipment	Used for insufflation during reptile coelioscopy
Sterile saline suspended above endoscopy table with intravenous drip line to a port on the operating sheath	Used for sterile saline infusion for cystoscopy, cloacoscopy, coelioscopy.
Flexible endoscopes	
Fiberscope, 2.5 mm × 100 cm working length, 1.2-mm instrument channel, 2-way tip deflection	Respiratory and gastrointestinal evaluations
Video bronchoscope, 5.9 mm × 61 cm working length, 2.2-mm instrument channel, 4 way tip deflection, suction and irrigation	Respiratory and gastrointestinal evaluations
Video gastroscope, 9 mm × 140 cm working length, 2.2-mm working channel, 4 way tip deflection, suction, insufflation and irrigation	gastrointestinal evaluations in large reptiles
Instruments for use with flexible endoscopes	
Grasping forceps, 1 mm × 120 cm Biopsy forceps, 1 mm × 120 cm	For 2.5-mm fiberscope
Biopsy forceps, round cupped jaws, diameter 1.7 mm, working length 120 cm Grasping forceps, alligator jaws, 1.7 mm × 120 cm	For 5.9-mm video bronchoscope
Biopsy forceps, round cupped jaws, diameter 1.8 mm, working length 180 cm Grasping forceps, alligator jaws, 1.8 mm × 180 cm	For 9.0-mm video gastroscope

Rigid telescope and instrument handling

The fact that most pet reptiles weigh < 2 kg requires careful control of the telescope with the eye-piece and camera supported using the superior hand, and the distal shaft of the telescope held by thumb and forefinger of the inferior hand. Handling the telescope in this fashion provides fine control without tremor; however, for a single surgeon to be able to use instruments a modified hold is required. In order to insert and manipulate instruments through the operating sheath it is necessary to change from a two-handed to a one-handed technique. The usual thumb and forefinger support of the telescope shaft is now adjusted so that the inferior hand takes the entire weight of the sheath-telescope-camera system. The thumb is slid up the shaft of the sheath, and the fingers are curled over the top to encircle the sheath. Now that the sheath is grasped in a fist-like grip with the sheath further supported by the thumb to prevent rotation, the superior hand can be removed to pick up and insert an instrument down the operating channel. This can only be performed using a correctly sheathed telescope; damage will otherwise occur.

Biopsy care and handling

It is interesting to note that various studies have concluded that endoscopic biopsies are more diagnostic than ultrasound-guided aspirates or biopsies.[3-5] Endoscopic biopsies are delicate and measure approximately 1 mm^3 when collected using 1.7-mm forceps. Handling and other histologic artifacts are reduced by gently dislodging the tissue from the biopsy cups into a small volume of sterile saline, before then decanting into a biopsy cassette or bag, and submitting in 10% neutral buffered formalin.[6] Picking the biopsy out of the instrument using a needle will cause severe damage, and even moistened cotton-tipped applicators have been shown to cause tissue alteration. Biopsies for microbiology are best submitted in sterile saline for immediate processing. Alternatively, if mailing to a laboratory, they should be submitted in appropriate transport media. For the submission of samples for toxicology or parasitology, it is wise to consult with the laboratory prior to sample collection.

Gastroenteroscopy

Sternal or dorsal recumbency with head and neck extended is preferred for stomatoscopy and gastroscopy. It is often helpful to have the reptile close to the table edge or positioned on a raised surface. The rigid telescope-sheath system can be used to examine the stomach of small lizards, and chelonians. Flexible endoscopes are required to gain access to the stomach for most snakes, and larger reptiles of other orders. Access to the anterior small intestine is more challenging in lizards and chelonians, than snakes. Air insufflation can be used to dilate the tract which provides good exposure for detecting gross lesions and foreign bodies. Warm saline irrigation provides superior clarity especially when examining for mucosal detail, although tracheal intubation is essential to avoid aspiration. Whether air or saline is used, it is important to gently dilate the tract as the endoscope is advanced to avoid damage and laceration to the wall. Foreign bodies can be readily retrieved and biopsy collection should be undertaken with care to avoid perforation of the intestinal tract, which is more likely in the thin-walled esophagus than the stomach or small intestine.

Cloacocoloscopy

While snakes can also be positioned in sternal or dorsal recumbency, chelonians and lizards are generally easier in dorsal. Endoscopy using warm saline irrigation promotes unparalleled examination of the proctodeum, urodeum, and coprodeum of the cloaca. Detailed appraisal of the distal colon, cloacal mucosa, urogenital papillae, oviductal openings in females, and (when present) the urethral opening and bladder. With practice, the endoscope can be directed through the urethra into the thin-walled bladder in chelonians, and some lizards. It is also possible to enter the distal oviducts of reproductively active females. Cloacoscopy has been used to remove shell and egg material from the distal oviducts and bladder, and as a guide for the exteriorization of partially prolapsed tissue and resection. Mucosal biopsies should be taken with care especially from the colon due to its thin nature, and the risk of perforation. More recently, Selleri presented a technique for neonate gender identification that centered on observing the gonads through the bladder wall via a cloacoscopic approach.[7] This technique has the advantage of avoiding entry into the coelomic cavity but, because examination is through a membrane, may not be as sensitive as direct coelioscopic visualization.

Coelioscopy

As reptiles lack a true abdomen, the term coelioscopy is preferred over laparoscopy. For a detailed discussion of generic laparoscopy methodologies the reader is referred to the dedicated literature as only reptile specifics will be highlighted here.[8-11] Laparoscopy has been shown to offer significant advantages over traditional surgical options, both in human and veterinary medicine. In particular, laparoscopy is, with practice, faster, less traumatic, less post-operative pain, and a more rapid return to normal function.[4,5,12-21] Until the advent of 2- and 3-mm human pediatric equipment, laparoscopy was limited to a single-entry system using the sheathed telescope; however, multiple-entry techniques are now possible and practical for animals over 500 g.[22]

Precise positioning for coelioscopy depends upon the reptile's body shape and confirmation, the structure(s) of interest, and the preferences of the endoscopist.[23] The entry site is aseptically prepared and draped, and adhesive clear plastic drapes permit better visualization of the patient. Insufflation is essential for lizards and snakes and although the shell prevents coelomic collapse can still be helpful in many chelonians. Typically CO_2 insufflation pressures of 3-5 mmHg are used, which are lower than those recommended for mammals (8-12 mmHg). In some situations, saline infusion may be preferred over gas. In the author's experience this is especially true when dealing with neonates and other very small reptiles, or when dealing with aquatic animals, where residual gas may have a negative effect on post-operative buoyancy.

Lizards

Dorsoventrad compressed lizards (eg, bearded dragons) are often positioned in dorsal recumbency, while laterally compressed lizards (eg, chameleons) are best placed in lateral. Round-bodied lizards (eg, green iguanas) can be placed in either lateral or dorsal recumbency, but lateral is generally preferred and allows easier access to both ventral and dorsal structures. Given the relatively small size of most lizards, a single paramedian or paralumbar entry point will usually permit examination of most, if not all, organ systems.

While anatomic differences exist between families, the green iguana serves as a useful model for saurian coelioscopy.[24,25] In the iguana, there is no significant difference in visualization of lung, liver, pancreas, small intestine, large intestine, ovary, oviduct, testis, epididymis, vas deferens, bladder, fat body, or kidney from either left or right approaches; however, a left lateral approach to the stomach, and spleen, and a right lateral approach to the gall bladder are preferred.[25] Right-handed surgeons will find that a left paralumbar approach is easier while left-handed surgeon will prefer a right paralumbar approach. Unless physical examination, diagnostic imaging or clinicopathology suggest an asymmetric lesion, the endoscopist's preference can prevail. The iguana is positioned in lateral recumbency, with the left pelvic limb taped caudad against the tail base. The entry area is bordered craniad by the ribs, dorsad by the spine, and caudad by the pelvic limb. Aseptically, a small skin incision is made in the center of the defined area. The skin and underlying muscle is grasped and elevated away from the coelomic viscera and the sheath and obturator are gently forced through the coelomic musculature and into the coelomic cavity. It is wise to temporarily cease artificial ventilation until the telescope has been introduced into the coelom, thereby reducing the possibility of damage to an inflated lung or viscus. The obturator is replaced by the telescope and the insufflation line attached to one of the sheath ports. By making a small skin incision and breach in the muscle, the sheath will be tight fitting and insufflation gas leakage will be minimal. Once a pneumocoelom has been created, it may be necessary to gently touch the tip of the telescope against a coelomic membrane to clean the terminal lens of condensation or tissue fluid. If there is fat or blood on the lens it is often necessary to remove the telescope from the sheath, clean with gauze moistened with sterile saline, and replace.

Upon entry via a left paralumbar approach in the iguanid lizard, the first organ to note is the large brown liver lying in the mid-ventral coelom. There are no diaphragmatic, post-pulmonary, or longitudinal membranes in iguanas; however, these membranes do exist to a greater or lesser extent in tegus, monitors, and certain other saurians. Minor perforation of these membranes by the telescope will not cause any harm but care is required to avoid perforating the underlying viscera. The stomach is located in the mid-coelom, and the spleen, an elongated dark red organ in iguanas, situated close behind. Careful examination medial to the stomach and spleen will reveal the splenic limb of the pancreas. The gonads are located just caudal to the spleen, either side of dorsal midline. Rigid endoscopy can be utilized to determine the gender of reptiles, even at a very early age. This is extremely useful in monomorphic species, or when involved in an endangered species breeding program that requires gender determination in neonates or juveniles.[26] Endoscopy also provides feedback on reproductive activity and disease. The testis is usually ovoid and smooth, and size may vary dramatically with season. The immature or inactive ovary appears as a cluster of small, fluid-filled follicles. As the ovary matures some of the follicles will enlarge and appear yellow to orange in color as they fill with yolk. Some species of lizards have thin post-hepatic and/or post-pulmonary membranes that may be pigmented and obscure the gonads. The membrane can be breached to allow better visualization by grasping and tearing with retrieval forceps, or incising using scissors. Care must be taken not to damage the follicles as leakage results in coelomitis.

Dorsal to the gonads are the adrenal glands which lie adjacent to the renal veins. The vas deferens of males and oviducts of females are also visible and can be followed caudad to the pelvic inlet. The kidneys of most lizards are situated in the caudodorsal coelom; however, those of the iguana are intrapelvic but can still be visualized. Moving ventrad from the pelvic inlet the bladder (if present) and fat body can be seen. A left paralumbar approach will reveal the small intestine and terminal colon; however, a right paralumbar approach will provide access to the gall bladder and caecum. The majority of the pancreas is easier to locate via a right approach, residing caudal to the liver and gall bladder and closely associated with the duodenum.

This approach to saurian coelioscopy has been validated, at least in the green iguana, and found to be both safe and effective.[25] In addition, endoscopic biopsies of liver and kidney have also been demonstrated to be safe and capable of producing samples of diagnostic quality.[27,28]

Snakes

Snakes are less commonly subjected to coelomic endoscopy because (i) they have more diffuse coelomic fat, (ii) it is impossible to examine all the major organs via a single entry point, (iii) insufflation is generally more difficult, and (iv) targeted coeliotomy can be readily performed along the entire length of the coelom. Nonetheless, if a targeted endoscopic approach to a single coelomic region is required, lateral entry between the ribs in larger snakes, and ventrolateral entry just medial to the ribs in smaller snakes are practical options.[23] A recently developed technique for liver biopsy entails access via the pulmonoscopy approach to the serpentine airsac. From within the airsac it is possible to visualize and biopsy the liver.

Chelonians

With the chelonian supported in lateral recumbency, the most useful coelioscopic approach is via the left or right prefemoral fossa.[23] Right-handed surgeons will generally find it easier to enter the left prefemoral fossa, and left-handed surgeons the right. Again, unless unilateral pathology exists, the surgeon's preference can prevail. A large bladder can hinder coelioscopy and chelonians should be encouraged to urinate prior to surgery. This can be achieved by bathing the animal, or by gentle digital stimulation of the cloaca prior to anesthetic induction.

The pelvic limb is retracted and taped caudad in order to expose the prefemoral fossa. In chelonians with a well-developed plastron hinge (eg, box turtles) it is wise to place a wedge between the caudal plastron and carapace to maintain exposure. Following aseptic preparation, a small (2-4 mm) craniocaudal skin incision is made in the center of the prefemoral fossa. The subcutaneous fat and connective tissues are bluntly dissected using hemostats down to the coelomic aponeurosis (formed by the broad tendinous portions of the transverse and oblique abdominal muscles) being careful to remain cranial to the sartorius muscle, and ventral to the iliacus muscle that lies between the femur and ventral surface of the ilium. The coelomic aponeurosis is penetrated by advancing hemostats (or the sheath and obturator) towards the head. Some force is required to breach of the coleomic membrane, and gain access to the coelomic cavity. An objective evaluation of coelioscopy in freshwater turtles (*Trachemys scripta* spp) has been published.[29] This study evaluated the ability of the 2.7-mm telescope to visualize the coelomic viscera of red-eared and yellow-bellied sliders, and collect diagnostic biopsies from liver and kidney. Ease of entry and organ visualization was satisfactory to excellent for all observed structures, except the spleen which could only be located from the right side.[30]

The dark red-brown liver is the most obvious organ to use for orientation. The stomach is most prominent on the left and, while potentially seen from either lateral, the pancreas and closely associated duodenum are easier to see from the right. The gall bladder is obvious and associated with the caudal edge of the right liver lobe, but the spleen is often hard to locate under the right liver lobe, close to the duodenum and pancreas. The ventral aspects of the lungs are located dorsad, and are most obvious in those species that lack a prominent post-pulmonary membrane. The colon, gonads, oviducts (in females), bladder and kidneys can be viewed from either side. The bladder is situated in the most dependent area, while the ovaries and oviducts of reproductively active females can occupy much of the coelom. The male testes (cream, yellow or brown in color) are situated in the caudodorsal coelom, and closely associated with the vasa deferentia, epididymides, adrenal glands, and retrocoelomic kidneys. Recently, coelioscopy using saline-infusion has been shown to be accurate for identifying gender in hatchling and neonate chelonians. A recent study in Chinese box turtles (*Cuora flavomarginata*) concluded that gonads could be visualized and sex easily identified in animals as small as 10 g.[26]

The kidneys are retrocoelomic, residing behind an often transparent coelomic membrane. When presented with a pigmented coelomic membrane, the close association with the testis (or immature ovary) can help indicate their location; however, in adult females where the ovary and suspensory ligament have enlarged and fallen away from the pigmented coelomic membrane it can be challenging to locate the kidneys unless they are grossly abnormal.[30] An extracoelomic approach to the chelonian kidney is also possible in those species with a more flattened carapace, and negates the need to enter the coelomic cavity. The sheathed endoscope is gently advanced from the same prefemoral skin incision in a caudodorsal direction between the coelomic aponeurosis and the broad iliacus muscle. A combination of gentle lateral movements of the telescope tip and intermittent insufflation are used to further separate the coelomic aponeurosis from the adjacent musculature to reveal the kidneys.

Pathology and Endoscopic Biopsy

In reptiles, a variety of pathologies may be appreciated endoscopically, often with surprisingly few clinicopathologic or discernible radiographic or ultrasonographic changes. Tissue samples can be easily harvested from most organs and, in general, any abnormal soft tissue structure. Liver disease may be focal or diffuse, and although the caudal edge is easiest to sample, biopsies can also be taken from the surface by first incising the serosal covering. A variety of renal diseases have also been described.[30,31] In lizards and snakes the lobulated kidney is easy to approach and biopsy directly; however, in chelonians its flattened retrocoelomic position first requires incision of the coelomic membrane via a coelioscopic approach. Endoscopic liver and kidney biopsies have been objectively assessed in freshwater turtles and green iguanas.[27-29] These studies have confirmed the ability of the 2.7-mm telescope to visualize the coelomic viscera and harvest samples from liver and kidney using the 1.7-mm biopsy forceps. No significant iatrogenic trauma was reported, and biopsy specimens were consistent with tissues collected at necropsy, and considered acceptable for histologic interpretation. Furthermore, the ability to diagnose renal disease using endoscopic biopsy has also been demonstrated for a variety of chelonian species.[30]

Post-operative Care, Complications, and Outcomes

Reptiles should be closely supervised until ambulatory, and be continuously maintained within their species-specific preferred optimum temperature zone. Meloxicam is relied upon post-operatively, although opiates and local anesthetics are useful as part of a balanced approach to analgesia. Continued fluid therapy is more important than an immediate return to feeding. Typically reptiles return to normal function and behaviors quicker following endoscopic procedures compared to more traditional, invasive approaches. Sutures are removed in 6-8 weeks.

The major complications encountered are typically associated with anesthesia and the advanced disease state of many presented reptiles. The importance of a thorough pre-operative evaluation and stabilization cannot be overemphasized. Ventilation, fluid support and temperature maintenance are critical. Minor hemorrhage following tissue biopsy is common but clinically insignificant. Most endoscopy issues are related to operator error until experience and ability have been gained. In general, the ability to examine reptiles internally and collect tissue samples greatly aids diagnosis and improves treatment success. To facilitate endoscopy caseload without compromising patient welfare, it is recommended that the surgeon obtains training and retain the option to convert to a traditional surgical approach if required.

Return to normal behaviors and weight gain are often the most useful indicators of improvement. Serial clinicopathology can also be useful if abnormalities were detected preoperatively. Serial endoscopic evaluations can be used to monitor patients, especially as some diseases (eg, hepatic lipidosis, glomerulonephrosis) may have a protracted course.

Training

A 3-day rigid endoscopy training course is offered December 1-3 2017 at the University of Georgia, and includes endoscopy and endosurgery of birds, reptiles and, for the first time, small mammals (incl otoscopy, stomatoscopy, laryngoscopy, laparoscopy and endoscopic ovariectomy). The course includes 12 hours of practical lab time during which veterinarians will practice skills taught in the classroom. Further details and on-line registration available at http://vet.uga.edu/ce/calendar. Course limited to 18 veterinarians.

Acknowledgments: The author would like to thank Karl Storz Veterinary Endoscopy and in particular Dr Chris Chamness, Mike Bateman, and Dan McMahon for supporting endoscopy research, development, and training at the University of Georgia. All equipment used by the author in his hospital was purchased and not donated, and he does not receive any payment from any endoscopy company.

References

1. Divers SJ. Diagnostic endoscopy In: Mader DR, Divers SJ, eds. *Current Therapy in Reptile Medicine and Surgery*. St Louis, MO: Elsevier;2014:154-178.

2. Divers SJ. Endoscope-assisted and endoscopic surgery In: Mader DR, Divers SJ, eds. *Current Therapy in Reptile Medicine and Surgery*. St Louis, MO: Elsevier;2014:179-196.

3. Roth L. Comparison of liver cytology and biopsy diagnoses in dogs and cats: 56 cases. *Vet Clin Pathol.* 2001;30:35-38.

4. Falcone RE, Wanamaker SR, Barnes F, et al. Laparoscopic vs. open wedge biopsy of the liver. *J Laparoendosc Surg.* 1993;3:325-329.

5. Rawlings CA, Diamond H, Howerth EW, et al. Diagnostic quality of percutaneous kidney biopsy specimens obtained with laparoscopy versus ultrasound guidance in dogs. *J Am Vet Med Assoc.* 2003;223:317-321.

6. Stahl SJ, Hernandez-Divers SJ, Cooper TL, et al. Evaluation of transcutaneous pulmonoscopy for examination and biopsy of the lungs of ball pythons and determination of preferred biopsy specimen handling and fixation procedures. *J Am Vet Med Assoc.* 2008;233:440-445.

7. Selleri P, Giorgi T, Melidone R, et al. Minimally-invasive endoscopic sex determination technique in juvenile marginated tortoises, *Testudo marginata. Proc Annu Conf Assoc Reptilian Amphib Vet.* 2011;148.

8. Tams TR, Rawlings CA. *Small Animal Endoscopy*. 3rd ed. St Louis, MO: Elsevier;2011.

9. McCarthy TC. *Veterinary Endoscopy for the Small Animal Practitioner*. St Louis, MO: Elsevier Saunders;2005.

10. Monnet E, Twedt DC. Laparoscopy. *Vet Clin North Am Small Anim Pract.* 2003;33:1147-1163.

11. Lhermette P, Sobel D. *BSAVA Manual of Canine and Feline Endoscopy and Endosurgery*. 1st ed. Gloucester, UK: BSAVA Publications;2008.

12. Golditch IM. Laparoscopy: advances and advantages. *Fertil Steril.* 1971;22:306-310.

13. Grauer GF, Twedt DC, Mero KN. Evaluation of laparoscopy for obtaining renal biopsy specimens from dogs and cats. *J Am Vet Med Assoc.* 1983;183:677-679.

14. Hancock RB, Lanz OI, Waldron DR, et al. Comparison of postoperative pain after ovariohysterectomy by harmonic scalpel-assisted laparoscopy compared with median celiotomy and ligation in dogs. *Vet Surg.* 2005;34:273-282.

15. Kehlet H. Surgical stress response: does endoscopic surgery confer an advantage? *World J Surg.* 1999;23: 801-807.

16. Lagares-Garcia JA, Bansidhar B, Moore RA. Benefits of laparoscopy in middle-aged patients. *Surg Endosc.* 2003;17:68-72.

17. Orlando R, Lirussi F, Okolicsanyi L. Laparoscopy and liver biopsy: further evidence that the two procedures improve the diagnosis of liver cirrhosis. A retrospective study of 1,003 consecutive examinations. *J Clin Gastroenterol.* 1990;12:47-52.

18. Reissman P, Gofrit O, Rivkind A. Exploratory laparoscopy: a crucial advantage of laparoscopic over standard appendectomy. *South Med J.* 1994;87:576.

19. Vander Velpen GC, Shimi SM, Cuschieri A. Diagnostic yield and management benefit of laparoscopy: a prospective audit. *Gut.* 1994;35:1617-1621.

20. Weickert U, Buttmann A, Jakobs R, et al. Diagnosis of liver cirrhosis: a comparison of modified ultrasound and laparoscopy in 100 consecutive patients. *J Clin Gastroenterol.* 2005;39:529-532.

21. Yu SY, Chiu JH, Loong CC, et al. Diagnostic laparoscopy: indication and benefit. *Zhonghua Yi Xue Za Zhi (Taipei).* 1997;59:158-163.

22. Hernandez-Divers SJ. Surgery: principles and techniques In: Raiti P, Girling S, eds. *Manual of Reptiles.* 2nd ed. Gloucester, UK: BSAVA Publications;2004:147-167.

23. Hernandez-Divers SJ, Hernandez-Divers SM, Wilson GH, et al. A review of reptile diagnostic coelioscopy. *J Herpetol Med Surg.* 2005;15:16-31.

24. Divers SJ. Lizard endoscopic techniques with particular regard to the green iguana (*Iguana iguana*). *Semin Avian Exot Pet Med.* 1999;8:122-129.

25. Hernandez-Divers SJ, Stahl S, Hernandez-Divers SM, et al. Coelomic endoscopy of the green iguana (*Iguana iguana*). *J Herpetol Med Surg.* 2004;14:10-18.

26. Hernandez-Divers SJ, Stahl SJ, Farrell R. Endoscopic gender identification of hatchling Chinese box turtles (*Cuora flavomarginata*) under local and general anesthesia. *J Am Vet Med Assoc.* 2009;234:800-804.

27. Hernandez-Divers SJ, Stahl SJ, McBride M, et al. Evaluation of an endoscopic liver biopsy technique in green iguanas. *J Am Vet Med Assoc.* 2007;230:1849-1853.

28. Hernandez-Divers SJ, Stahl SJ, Stedman NL, et al. Renal evaluation in the healthy green iguana (*Iguana iguana*): assessment of plasma biochemistry, glomerular filtration rate, and endoscopic biopsy. *J Zoo Wildl Med.* 2005;36:155-168.

29. Divers SJ, Stahl SJ, Camus A. Evaluation of diagnostic coelioscopy including liver and kidney biopsy in freshwater turtles (*Trachemys scripta*). *J Zoo Wildl Med.* 2010;41:677-687.

30. Hernandez-Divers SJ. Endoscopic renal evaluation and biopsy of Chelonia. *Vet Rec.* 2004;154:73-80.

31. Hernandez-Divers SJ, Innis C. Renal disease in reptiles: diagnosis and clinical management. In: Mader D, ed. *Reptile Medicine and Surgery.* St. Louis, MO: Elsevier Saunders;2006:878-892.

Gap in pagination due to unavailable paper.

Pages 248-319

Section 21

ARAV Bonus Master Class

Anneliese Strunk, DVM, DABVP (Avian Practice);
Mike Miller, DVM;
Karina Mathes, Dr med vet, Cert Spec in Reptiles and Amphibians, DipECZM (Herpetology)
Session Chairs

Reptile and Amphibian Nutrition

Thomas H. Boyer, DVM, DABVP (Reptile and Amphibian Practice), and
Peter Scott

Session #2131

Affiliation: The Pet Hospital of Penasquitos, 9888-F Carmel Mountain Road, San Diego, CA 92129, USA.

Abstract: Nutrition is the leading cause of disease for reptiles and amphibians. Water is the most essential nutrient for both groups. Veterinarians need to be able to evaluate complete manufactured diets by reading the product label, converting the guaranteed analysis to a dry matter basis, looking at calcium content and, if possible, calculating metabolizable energy and protein to energy ratios. Likewise, multivitamins and calcium supplements have to be evaluated to make sure they are appropriate, as many are not. Insectivores require a variety of gut loaded and calcium dusted insects. Herbivores are a challenge to feed salad-based diets, complete manufactured diets, grasses and hays are compared. It is easier to feed whole vertebrates to carnivores, although fish as a diet source can be problematic, if not handled appropriately. Omnivores, including box turtles, aquatic turtles and bearded dragons also are discussed.

Masterclass adapted from: Nutrition and Nutritional Diseases. In: Divers S, Stahl S, Mader D, eds. *Mader's Reptile and Amphibian Medicine and Surgery.* 3rd ed. St Louis, MO: Elsevier Saunders; In press.

Introduction

Nutrition is the leading cause of disease in reptiles and amphibians. Both owners and veterinarians often know little about herptile nutrition, and information. Information from the internet, pet stores, and outdated literature is often misguided. Chronic nutritional diseases remain common, including nutritional secondary hyperparathyroidism (NSHP), hepatic lipidosis, protein deficiency, hypovitaminosis A, hypervitaminosis A, pyramidal shell growth, renal disease, urocystoliths, thiamine deficiency, vitamin E/selenium deficiency, steatitis, corneal lipidosis, obesity, loose stools, and starvation. The goal of this lecture is to educate veterinarians so that they can provide sound nutritional advice to reptile and amphibian keepers.

Water

All animals need water. It is the most essential nutrient, with enormous diversity in methods utilized to obtain it. Mammals can survive loss of most fat and glycogen reserves, and half of their bodies' protein, but a 10-15% water loss is life threatening. Water remains of critical importance to reptiles, even desert species, which can withstand greater losses, but still require water. Three sources of water are available to animals: dietary water (in food), metabolic water (formed from cellular metabolism) and environmental water. Few animals are able to survive without environmental water, and those that do are generally herbivores consuming plants with a high water content. Their dependence, or independence, on environmental water is not well understood.[1] Many desert species can behaviorally conserve water by retreating to a humid underground burrow, which may not be available in a cage or yard with low relative humidity. Some reptiles are not adept at drinking from water bowls. All reptiles and amphibians need environmental water, whether misted, sprayed, dripped or provided as free-standing water.

Complete Manufactured Diets

The main advantage of manufactured diets is they can contain all of nutrients for a given species (assuming we know what they are). The disadvantages are that if any nutrients are omitted or included at inappropriate levels, the consequences can be substantial.[2] No regulatory agencies oversee reptile diets or advertising claims, so veterinarians should be cognizant of a wide variation in quality and be prepared to evaluate the diets themselves. The data and research to formulate complete herp diets is not currently available. Some use National Research Council (NRC) recommendations for other species (such as laboratory rats and domestic carnivores) as a starting point, or known nutritional studies from herps, where they are available. Some complete diets are from feed manufacturers with PhD nutritionists or veterinarians on staff to provide professional input, while others are from pet suppliers with little expertise in nutrition.

Reading a product label

Protein is the most costly component, and therefore manufactured diets will often use less expensive plant proteins in combination with smaller quantities of mammalian or fish proteins. Examining ingredient lists may help estimate digestibility since plant materials are generally less digestible by carnivores. Reviews of how to 'read' labels are available.[3,4]

Guaranteed analysis

If ingredients are listed without a guaranteed analysis, then readers are advised to be skeptical. The more nutritional information provided, the better. The water content of foods varies, so for direct comparison diets are compared on a dry matter basis (DMB), with water removed, and not by the nutrient levels on the label (which are as fed, AF).

Metabolizable energy

Another way to look at foods is by metabolizable energy (ME) or kcal ME/cup. In general, juvenile and carnivorous reptiles require higher dietary ME than adults.

Protein

A protein:energy ratio (P:E ratio) can also be calculated for a diet. In general the P:E ratio needed by an animal changes through its life, with a higher proportion of protein to energy needed as animals are growing. It also varies depending on activity and reproductive status. The consumption of excess calories resulting in a maximal growth rate is not compatible with optimal skeletal development and may result in skeletal deformities in juvenile chelonians and power-fed snakes. Animals short of energy will use protein as an energy source, whereas a physiological excess of protein tends to be stored as fats. The metabolism of excess protein in this way can lead to excess nitrogenous waste excretion

Calcium

Healthy Ca to P ratios for most vertebrates are generally 1-2:1, but may be higher for growing herps. For example, rapidly growing juvenile saltwater crocodiles, *Crocodylus porosus*, select a diet with an overall Ca:P ratio of 6.7 to 1, primarily because of the calcium-rich crustacean shells. Adult saltwater crocodiles consume more mammals and fish, but less crustaceans, and consequently the Ca:P ratio of their diet drops to 1.85 to 1.[5] Desert tortoises diets were found to have 3.2-5.8:1 C:P ratio.

Multivitamins and Calcium Supplements

Multivitamins can be broken down into those with or without Ca. This author discourages MV's with Ca because they do not contain enough Ca to prevent NSHP and, if used on a daily basis, can lead to oversupplementation of fat soluble vitamins A and D, or trace minerals, such as copper, iron or zinc. For most animals, the presumed maximal safe levels of vitamins A and D_3 for long term feeding (>60 days) is 4 to 10 times the recognized dietary requirement.[6] Unfortunately for herps, we still do not know their dietary requirements for either A or D, nor any other vitamins. In iguanas, the risk of over supplementation of vitamin A was thought to be especially great in multivitamins with vitamin A levels > 1,100 IU/g.[2] The estimated daily vitamin A requirements of green iguanas was 56 IU/kg and chronic intake of vitamin A 2-3 times the recommended dietary level of 8,000 IU/kg feed could produce hypervitaminosis A.[2] Using human formulated products should be done with great care, as it may be impossible to subdivide a tablet into small enough portions for safe dosing.[2] If multivitamins are used, the amount provided should be restricted to that amount that supplies the daily requirement for vitamin A, but at this level many supplements are poor sources of other vitamins and trace minerals.[2] This author has had empirical success recommending multivitamins once or twice a month for diets lacking vitamin A, and no multivitamins if commercial gut loading diets or dry formulated feeds with vitamin A are used. Vitamin deficiencies typically arise from a total lack of the vitamin in the diet, small amounts of the vitamin, with a balanced diet, generally avoid deficiencies.

Multivitamins without Ca should be evaluated to see if they are true multivitamins (all vitamins A, B, C, D, E, and K as well as trace minerals). Insectivores, such as chameleons, leopard geckos, toads and some frogs, need vitamin A in their diet because many cannot convert β-carotene precursors to vitamin A. All herbivores seem capable of vitamin A synthesis. Most carnivores typically avoid vitamin A deficiency because liver in whole animal diets is an excellent source of vitamin A. An informal survey of popular reptile multivitamins showed about a third contained no vitamin A, but instead list β-carotene, and half are not true multivitamins. This could be an issue for insectivores if not fed properly gut loaded insects, or for Emydid turtles, if their diet were vitamin A deficient. The ratio of A:D:E should be roughly 100:10:1. Essential amino acid supplementation should not be necessary if insects, whole animals, or nutritionally complete commercial diets are fed, all of which have adequate essential amino acids.

The vitamin D content must also be considered, reptiles are thought to use vitamin D_3 (cholecalciferol), not D_2 (ergocalciferol). Vitamin D is a fat-soluble vitamin or hormone, and excess may be retained in fat. Large doses of vitamin D can be toxic (hence its widespread usage as a rodenticide), although these doses far exceed levels used in most supplements. Not all reptiles (especially *Iguana* and *Cyclura* sp.) seem to be able to efficiently utilize dietary vitamin D.[2,7] If adequate UVB light exposure is available there is less need for dietary vitamin D_3 in Ca supplements, especially if multivitamins or fortified foods containing vitamin D_3 are already in use.

Calcium supplementation is needed on a regular basis (daily to every other day) for all growing herps not consuming whole vertebrates regularly. Once adults have fully formed skeletons, Ca demands decrease, but there is still a need, especially in reproductive females. Most calcium salts are better absorbed with food, and help to competitively inhibit P absorption. Ideally, Ca supplements should be independent of multivitamins (other than vitamin D), and contain little to no P. Most insects have a 1:5 -10 Ca:P ratio, therefore a Ca supplement with a 2:1 Ca:P ratio (common with bone meal products) will never achieve a positive Ca:P ratio until the mass of the supplement is greater than the mass of the insect. All herp Ca supplement should be P free. Beware of similar sounding products which makes distinguishing Ca powders more confusing, have owners bring supplements to their office visit so they can be carefully evaluated.

The type of Ca used is also important as different kinds of Ca salts have different amounts of elemental Ca available for absorption and utilization. Ca carbonate is the best. One gram of Ca carbonate has 40%, or 400 mg elemental Ca, available for absorption, followed by 1 gram of Ca citrate with 211 mg. You would have to use almost three times as much Ca lactate to equal Ca carbonate in elemental Ca. How well the product adheres to the insect depends on electrostatic properties and the fineness of grind. Finer particles adhere better than coarser particles. Also heavily dusting salads adds a mildly salty, sour, or chalky taste (for humans), which also appears to reduce palatability for herps.

Nutritional Biology

Although some particularly interesting exceptions exist, caecilians generally feed on earthworms and other invertebrates, frogs and salamanders feed almost exclusively on insects (at least as adults, tadpoles are detritivores), crocodilians feed largely on other vertebrates, turtles feed on a combination of plants and animals, and squamates feed largely on invertebrates or vertebrates. Herbivory is uncommon and is generally the exception within groups, rather than the rule, except for Testudinidae, Iguanidae, and Leiolepidinae (Butterfly agamids).[8] Adult amphibians and reptiles do not necessarily eat the same prey as larvae or juveniles, ontogenetic shifts in diet are probably common.[8] On a very simplistic level, reptile and amphibian diets can be categorized as insectivores, carnivores, herbivores, or omnivores. Many reptiles have specialized dietary needs that require extensive appreciation of natural history.

Insectivores

This group includes many lizards, most amphibians, some turtles, and a few snakes. Many obligate carnivorous adults may be insectivores as growing juveniles. Insectivory is a broad term which also includes consumption of arachnids, annelids and crustaceans.[9] Nutritional requirements of most insectivorous reptiles are unknown so nutritionists often extrapolate from better known species, such as the rat, or other carnivores, that have known NRC requirements.[10-12] Deficiencies by nutrient (percent of samples deficient versus NRC rat growth requirements on a DMB) for fasted adult and nymphal house crickets (*Acheta domesticus*), superworms (*Zophobas morio* larvae), larval and adult mealworms (*Tenebrio molitor*), giant mealworms (*Tenebrio molitor* mealworm larvae treated with juvenile hormone to prevent them from molting into adults), waxworm larvae (*Galleria mellonella*), silkworm larvae (*Bombyx mori*) and earthworms (*Lumbricus terresstris*), were as follows: calcium (100%), vitamin D3 (100%), vitamin A (89%), vitamin B_{12} (75%), thiamin (63%), vitamin E (50%), iodine (44%), manganese (22%), methionine-cystine (22%) and sodium (11%).[11] Simply put, insects are Ca and multivitamin deficient. All invertebrates are a good source of protein and meet or exceed NRC requirements for growth in rats.[11] Some of the protein nitrogen may be chitin: some insectivores have chitinases, while others rely on gut bacterial chitinases.[9] Invertebrates contain little or no carbohydrates except silkworms fed a high (60%) carbohydrate diet.[11] Larval insects (except silkworm larvae) have high fat content (35-60% DMB), especially mealworms, waxworms and butterworms (*Chilecomadia moorei*).[12] If these larvae constitute a substantial portion of the diet, this may lead to disproportionally high fat content.[9] Insects contain significant quantities of fiber, highest in mealworms and lowest in silkworms, earthworms have almost no fiber content.[11] Invertebrates appear to be a good source of essential amino acids, except for superworms and waxworms, which are limited only in methionine + cysteine,[11] and adequate sources of trace minerals copper, iron, manganese and zinc.[10] Calcium and vitamin A and D levels of insects are below the NRC requirements for rat growth[11,12] and for carnivores.[10] Consequently Ca and vitamin A deficiencies are the most common nutritional diseases of both reptiles and amphibians. All analyzed insects are Ca deficient, with few exceptions (such as black soldier fly, *Hermetia illucens*). Earthworms, night crawlers (*Lumbricus terrestris*), and terrestrial crustaceans (Isopods), such as the wood louse (*Porcellio scaber*) and pill bugs, or roly-poly's (*Armadillidium* sp.), all have positive Ca:P ratios.[11,12] Crustacean shells are good sources of calcium carbonate.[13] Compared to NRC requirements for growing rats, most invertebrates contain sufficient vitamin E (except mealworms and superworms) and trace minerals, except superworms, giant mealworm larvae, waxworms, and silkworms, which are deficient in iodine.[11] Vitamin B_{12} is below the NRC requirements for growing rats for all larval insects. Thiamine is below the NRC requirements for growing rats in crickets, superworms, giant and adult mealworm larvae.

From this discussion it is obvious that commercially available invertebrates and annelids are nutritionally incomplete.[9] Feeding herps store bought crickets and mealworms without supplementation will result in calcium and/or vitamin A deficiencies, and may induce obesity and hepatic lipidosis. Improving insectivore nutrition is a simple triad of feeding a nutritionally complete calcium rich diet to insects, dusting insects with calcium, and less frequently multivitamins, and feeding a wide variety of insects, crustaceans, annelids and rodent pups. Both gut loading and dusting are recommended, gut loading provides more balanced nutrition and boosts Ca, and dusting provides more Ca.

Gut loading insects: Originally, feeding a Ca-rich diet to improve the insect's Ca content was called gut loading.[14] Early studies in feeder insects demonstrated that feeding high Ca diets (≥ 8% Ca DM) increased the Ca content and developed a positive Ca:P ratio of feeder insects after 24 hours (mealworms), 48 hours (crickets), or 72 hours (waxworms).[14-16] Higher Ca levels (12%) result in poor viability of mealworms and Ca levels begin to decline after a week of high Ca diets.[17] Unfortunately, over time, the term gut loading has degenerated into just feeding crickets, often with foods rich in β-carotene (even though crickets and some insectivores do not convert β-carotene to vitamin A).[18] For instance, feeding mealworms and crickets a 3% Ca DM diet with a 2.8 Ca:P ratio for 72 hours resulted in a mean Ca:P ratios of 0.19 for mealworms and 0.26 for crickets (ie, negative Ca:P ratios).[17] Not all cricket diets designed to boost Ca content are equal. In one study, 3 out of 4 commercial calcium-fortified dry diets designed for feeder crickets, contained no more calcium than unfortified diets.[19] Calcium-fortified high-moisture cricket waters or high-moisture foods (gel water cubes) are also ineffective at increasing calcium content.[20] A typical cube has 600 ppm Ca or 0.06% Ca, about twice the content of tap water. One study noted that as fed gel water cube labels claimed 0.1-0.12% Ca, but when analyzed were found to have only 0.01% Ca, and caused some mealworms to asphyxiate.[21] Another problem with gel water cubes is that insects, as well as reptiles, will consume them preferentially over high Ca diets. Instead of gel water cubes, offer water, such as a water soaked cotton ball in a bottle cap (for crickets and roaches) or damp paper towel (for mealworms or superworms). One of the reasons cubes became popular is that insects drown in standing water. Another study found that by day 3 mealworms on a Ca-fortified insect diet and offered either apple slices or gel water cubes, had mean negative Ca:P ratios of 0.62 and 0.91, respectively.[21] Do not offer fruit, vegetables, or gel water cubes, as a water source to feeder insects. Assume store-bought insects are nutritionally inadequate at purchase. Mealworms are typically shipped and sold in 100% wheat bran, with 0.14% Ca and a Ca:P ratio of 0.12, or even worse, inedible saw dust.[17] Owners should purchase a high-quality gut-loading diet or veterinarians should buy, stock and supply good products. Food products should be appropriately stored in a cool dry area, refrigerator, or freezer, and discarded after expiration dates of 6 to 12 months.[22,23] Crickets, mealworms, superworms and Dubia cockroaches (*Blaptica dubia*) can be maintained on cricket high Ca (≥ 8% Ca DM) diets, even though these diets are not intended as maintenance diets. Newer commercial gut loading diets can produce a positive Ca:P ratio as quickly as 12 hours, compared to experimental diets which required several days. Be sure to use a Ca-enriched diet (such as Mazuri Hi Calcium Gut Loading Diet, 8-9% Ca) and not a maintenance diet (such as Mazuri Cricket Diet, 1% Ca). Remember, water is the most essential nutrient, and insects will die rapidly without it (not from feeding a Ca-enriched diet). Feeder insects should not be refrigerated; they should be maintained at their preferred optimum temperature zone, when in doubt, around 27 °C (80 °F). Refrigeration merely extends shelf life without essential nutritional supplementation. Feeding starved insects will result in a starved reptile, even though the reptile is still eating.

Many of the gut loading studies produced variable results on final nutrient content (especially Ca:P) because of variations in insect age, developmental stage, size or sex, temperature, water sources, or lack of water, humidity, and, of course, palatability.[22,23] The most important is developmental stage. All insects grow through various life stages by shedding their exoskeleton. The time between sheds is called an instar, or growth period. Crickets begin feeding about 6 hours after shedding, then feed well until about mid-instar, at which point they have filled their exoskeleton and reached their critical weight to get through molting. At this stage, they reduce feeding until after the next molt. Instars can last days (at 35 °C or 95 °F) to weeks (at 28 °C or 82 °F), last longer in older crickets, and have obvious implications on how well the insect can be gut loaded.[22,23] The best way to minimize these variables is to gut load all insects, with the hope that short term deficiencies will be corrected over the long term.

Dusting insects: An additional essential method of improving Ca, or vitamin content, is dusting insects. To dust, insects and the Ca, or multivitamin powder, are placed in a plastic bag, or deep plastic cup, and lightly shaken or swirled, until the insects are coated. This method can produce variable results depending on the type of calcium or multivitamin used (eg, particle size and electrostatic properties). One study found crickets would groom off 50% of either multi-vitamin/mineral or calcium/vitamin D powder within 90 seconds of dusting.[24] A larger study involving two cricket species and multiple life stages, found dusting with Ca boosted Ca:P ratio above 1:1 for 5.5 hours (endpoint of study).[25] There were no statistically significant differences between the dusting powders, but there were differences between cricket species and life stages. Dusting insects in environments with high humidity is not suitable to enhance nutrient content, nor for aquatic insectivores, unless they take insects off the water surface.[26]

Insect variety: The third vital aspect of feeding insectivores is to offer a wide variety of insects. Insectivores naturally consume a diverse diet. Store bought insects such as crickets, waxworms, mealworms, super mealworms and Dubia cockroaches, should be supplemented with commercially available silkworms, black soldier fly larvae (Phoenix worms), tomato horn worms, bean beetles (*Callosobruchus maculates*), fruit flies (*Drosophila melanogaster, D hydei*), springtails (Collembolands), and wood lice (European wood lice, *Porcellio scaber*, dwarf tropical wood lice, *Trichorhina tomentosa*), as well as wild caught seasonally available insects such as moths, cicadas, flies, grasshoppers, bees (remove stingers), cockroaches, and crustaceans, such as wood lice and sow bugs (pill bugs or roly polys). Insects are easily collected at night around lights, with funnel traps or by children, and pill bugs accumulate in damp soil under a piece of shaded plywood. Owners often are overly concerned about pesticide hazards but this is rarely ever a problem. Fireflies (*Photinus* sp.) contain toxic lucibufagins, deaths have been reported in reptiles and amphibians after consumption.[27] Many insectivores can be trained or naturally take baby mice and annelids, which are excellent dietary supplements. Recommend that owners feed as wide a variety of invertebrates as possible, not just 2 or 3 varieties.

Carnivores

This group includes all snakes, crocodilians, and adult amphibians, and most aquatic turtles and lizards. Carnivores require high levels of protein (25-60%) and fat (30-60%), and low levels of carbohydrates (< 10% DM).[28,29]

Snakes: One reason snakes remain popular is that rodents are a completely balanced diet, easily procured, and thus nutrition related diseases are less common. Nutritionists consider whole vertebrate animals nutritionally complete foods that do not require supplementation.[30,31] Most pet stores, as well as numerous on-line vendors, sell frozen mice and rats, in all sizes. Veterinarians should be aware of different rodent life stages as owners often ask what size rodent to feed. Pinkies are neonatal rodents (*Mus musculus* and *Rattus norvegicus*) without fur and have been found to have an almost even Ca:P ratio (slightly negative to slightly positive). After 5-9 days mice develop fur, become fuzzies and from then on have a positive Ca to P ratio. After 2-2.5 weeks, mice open their eyes, and are known as hoppers (because of their propensity to jump when startled). By 21-25 days they wean, and are known as weanlings or small adults, then on to large, and extra-large adults. Rodents and other whole animal prey should be frozen in moisture impermeable plastic or vacuum packed to reduce dehydration during storage.[31] Frozen whole animals should be completely thawed and warmed before feeding to prevent a bacterial bloom in the prey's gut. This also occurs commonly if rodents are allowed to thaw over several hours at room temperature. Feeding cold or partly frozen prey is not recommended as it can cause gastrointestinal dysfunction and slowed digestion. For the die-hard reptile breeder still producing their own rodents it is recommended to feed commercial pelleted rodent diets, not seed mixes or dog food (which can pass on nutritional deficiencies to the predator). Feeding live prey, although legal in the United States, is ethically unacceptable in most countries and should be avoided. Prey should be humanely euthanatized (cervical dislocation, or gassed with CO_2) prior to feeding.[32] Serious trauma can be inflicted by a live rodent on a reptile. Given the choice, boid and colubrid snakes prefer thawed mice over freshly killed ones.[33] Rabbits, hamsters, gerbils, chicks, lizards, snakes, frogs, toads or fish are also appropriate food items. Muscle meat and viscera, lack bone, and therefore have little Ca. For example, Ca:P ratios of chicken meat (0.08:1), hamburger (0.06:1), chicken liver (0.14:1) and gizzards (0.1:1) are all inadequate for bone growth or maintenance.

Hatchling snakes often refuse pinkies and several sequential tricks can be used to induce feeding. Handling should be curtailed, and at least one hide box should be present in the enclosure. A pink mouse can be left in the cage overnight. Washing, to remove all rodent scent, and/or then scenting a pink mouse with the preferred natural prey (such as lizards for king snakes and milk snakes [*Lampropeltis* spp.], toads for hognoses [*Heterodon* spp.] or fish for garter, ribbon [*Thamnophis* spp.] or water snakes [*Nerodia* spp.]). Scenting involves rubbing, or attaching a small amount of skin, or limb, from the preferred prey, onto the pink mouse, especially around the head. "Braining" also works very well, even in snakes that have steadfastly refused pinkies. Brain material is forced out the pink mouse's nostrils or through a skull incision, onto the snout, and presented close to the snake, or left in the cage overnight. Tongue flicking by the snake of the pinky brain seems to be a super optimal stimulus releaser for feeding.

Fish: Fish is often used as a food for snakes, turtles and crocodilians. Whole fish are relatively good sources of most nutrients except for thiamine and vitamin E.[34] Thiaminases, enzymes that destroy or inactivate thiamin, are present in many marine animals, such as clams, herring, smelt and mackerel.[34] The recommended supplementation is 25-30 mg thiamine per kg of wet fish. Trout and most freshwater fish do not contain thiaminases and do not require thiamin supplementation. Vitamin E is abundant in fresh fish, but oxidation of fish lipids while frozen can deplete vitamin E content, even after a few weeks of freezing, and frozen fish should be used within 4 to 6 months.[30,34] The recommended level of supplementation is 100 IU of vitamin E per kg of wet fish. Many freshwater reptiles will not eat saltwater fish. The types of fish fed should be varied, and many piscivores will also eat rodents, snails, slugs, earthworms, shrimp, krill (frozen, not freeze dried) or pelleted commercial diets. Aquatic turtles do well on fish-based aquatic turtle pellets, earthworms, waxworms, mealworms, crickets, whole fish (freshly killed guppies or minnows, chopped trout and freshwater smelt), whole shrimp and pinkies or chopped skinned mice. Goldfish are not recommended due to frequent infection with *Mycobacteria.*

Crocodilians: Nutritional studies on salt water crocodiles, *Crocodylus porusus*, reveal average diets to be 70% water, 13-15% protein, 3-4% fat and with a Ca:P ratio of 2-7:1 which are similar to other crocodilians.[5, 35] Crocodilians are generalist opportunistic carnivores which eat a wide variety of prey, with the exception of Gharials, *Gavialis gangeticus,* which are fish specialists. Hatchlings feed mostly on small prey items such as insects, spiders, crustaceans, fish, amphibians; adults feed mainly on larger prey (fish, crustaceans, mammals, reptiles, and birds). Crocodiles are ambush predators and active scavengers, but will not take putrid prey, can be cannibalistic, and some occasionally even feed on fruits.[5] Small captive crocodilians can be fed gut loaded insects, rodents, and whole fish, or shrimp. Larger crocodilians can be fed larger prey such as crustaceans, fish, rodents, rabbits, chickens, pigs, nutria, cats, dogs, and sheep. Crocodilian farms often utilize waste food, or byproducts, unfit for human consumption, which works if animals are culled young, but can be lethal to long term zoo captives, which is unacceptable.[35] Dead rancid fish will can cause steatitis. As a general rule, juveniles should be offered food 2 to 3 times/week, sub adults 1 to 2 times/week, and adults once a week to every 2 weeks.[35] All crocodilians can be fed commercial crocodilian extruded biscuits.

Herbivores

Herbivores average dietary composition is 15-35% protein, less than 10% fat, 50-75% carbohydrates and 15-40% crude fiber (DM).[28,29] Herbivores are the greatest challenge to feed correctly. Extant herbivorous reptiles are found in the Cheloniidae, Testudinae, and some Agamidae and Iguanidae. Herbivory in lizards is limited to only about 50 species, and folivory is unique to the green iguana, *Iguana iguana*.[2,36,37] Herbivorous species are popular and a large portion of the reptile veterinarian's caseload. Herbivores are typically generalized grazers. Some species are only herbivorous as adults and insectivores, carnivores or omnivores as juveniles, such as the bearded dragon, and Emydid turtles, including *Chrysemys*, *Pseudemys*, *Trachemys* and *Terrapene*.[38] All herbivores require UVB light, green iguanas and rock iguanas (*Cyclura* sp.) seem to utilize dietary vitamin D relatively inefficiently, they seem far more dependent on UVB radiation.[2,7]

A common misconception is that herbivores will select a balanced diet from the foods offered. While this may be true for free-ranging herbivores, this is not the case in captivity. Complex natural diets of herbivores are only slowly beginning to be appreciated. For example, Sonoran desert tortoises (*Gopherus morafakai*) consume well over 200 plant species, and this dietary diversity is probably true of many tortoise species.[39] Plants vary seasonally, annually (dependent on rainfall), phenologically (freshest plants eaten first, dry senescent plants last) and between habitats. There is much intraspecies and interspecies variability, even between closely related species. One researcher commented that the abundance and diversity of annual plants varied so much from year to year, and site to site, that it was difficult, if not misleading, to characterize an average desert tortoise diet.[40] Nonetheless, captive desert tortoise diets should attempt to emulate the nutrient composition of wild diets, with a goal of ≤ 15% crude protein, 2-4% fat, and 25-30% fiber (%DM).[41,42] Captive herbivore nutrition is more of an art than a science, primarily because there is a scarcity of nutritional science to what wild herbivorous reptiles eat. Nutritional studies are difficult and tedious. Consider, for example, studying the green iguana. Green Iguanas are tough to observe in the field because they are arboreal folivores and very wary of humans. They are also inactive 90% of the time[43] and spend little time foraging (0.6% of total observation time in one study).[38] Once you observe them foraging, plants consumed must be collected, from trees, and nutritionally analyzed; however, one must be careful to collect exactly what they consumed. In another study, iguanas never ate mature leaves with ≤20% DM crude protein content, but always selected younger leaves with higher crude protein contents of ≥ 25% DM.[44] It is therefore understandable that we still know so little about the nutrient composition of wild iguana diets. Nutritionists, working on captive propagation of green iguanas for decades, have produced some valuable estimates.[2] For green iguanas, the recommended DM nutrient levels are 26% gross protein for hatchlings to 6 months, 22% gross protein after 6 months, 15 to 17% gross protein for non-reproductive adults, as well as 3% fat, 6 to 10% crude fiber and a 2:1 Ca:P ratio (% DM).

Salad-type diets: Salad-type diets are problematic in that commercial fruits and vegetables produced for humans typically are too low in Ca, fiber, trace minerals and some vitamins, too high or too low in protein, and too high in carbohydrates for herbivorous reptiles.[2] No surprise that captive herbivores have calcium deficiencies, difficulty laying eggs, bladder stones, renal disease, and loose stools. In southern California, chronic nutritional disease is endemic in tortoises fed back yard grasses, weeds, grocery store fruits and vegetables. Most suffer from chronic protein, Ca and fiber deficiency, and carbohydrate excess, with many dying from hepatic lipidosis.[45] Captive iguanas typically die from Ca deficiency, hypovitaminosis D (metastatic mineralization), or renal failure. It is difficult to provide a balanced range of necessary nutrients with greens, vegetables and fruits. That is not to say that they have no place in in herbivorous reptile diets, but, by themselves, are often poor alternatives to wild foods.[2] Another problem with salads is that animals can self-select, such as with fruit, and individual food items are generally nutritionally incomplete. The USDA nutrient database (https://ndb.nal.usda.gov/ndb/) is an on-line, searchable database that enables veterinarians to obtain complete nutritional information on any human food item.

Attempts to compensate for nutritional inadequacies in salad diets by supplementing with vitamin-mineral supplements is difficult without knowing what they need, how much they need, or how much they are consuming and doesn't correct other nutrient inadequacies or excesses. This also applies to natural foods planted for browse in the herbivores exhibit. In the wild, they consume dozens to hundreds of species, although they may concentrate on a few. Most backyards or exhibits simply are too space limited to support a wide variety of species and herbivores often overgraze whatever is in their exhibit. This is not to say that native palatable species should not be offered, but rather that it is hard to provide a nutritionally complete diet in such a manner.

There is a difficult trade-off to be made between trying to mimic a reasonably normal diet whilst accepting that the actual components are nothing like those encountered in the wild, against using a totally abnormal commercial diet based on in-house extrapolation with little to no validation. The former presumably fills more of the non-nutritional behavioral needs of the animal, which will not be experienced at eating uniform pellets. Some prefer to attempt to simulate normality in the types of food items presented often with the choices dependent on availability of suitable natural or commercial foodstuffs. Others believe providing a high quality commercial pelleted diet as a large portion of the diet probably offers a superior dietary alternative to what the typical owner can accomplish.

Grasses and hays: Most grasses and hays, except for Kentucky bluegrass and alfalfa, approach or meet the nutrient profile for desert tortoises, except some are low in Ca, and are recommended.[41] Clover (*Trifolium* spp.) and alfalfa (*Medicago sativa*) have more Ca but are also higher in protein. Alfalfa is a nitrogen fixing legume hay that has higher crude protein than grass hays.[42,49]

Complete manufactured diets: Commercial diets were compared to nutritional analysis of the most frequently consumed plant species reported from 4 studies of free-ranging desert tortoises[41,46-48] and recommended nutrient levels for green iguanas.[2,46-48] On a DM basis, there was a great deal of variation in these diets, and some are better than others in approaching the free-ranging nutrient profiles. Overall, they appear better than greens, fruits or vegetables; however, a diet for juvenile iguanas with a 13% protein DM, is not only inappropriate for juveniles, but for adult iguanas as well. Only 2 of the tortoise diets had the recommended 3.2-5.8 Ca:P ratio, so additional Ca supplementation may be needed, and all tortoise diets were high in carbohydrates.

Getting tortoises to eat hay and commercial foods can be difficult. Much like a dog content with table food, a tortoise will eat pellets and hay once hungry enough, but only if better tasting foods are not available. Feeding less frequently will encourage the reptile to try new foods. Use good quality hays, not stems or stalks, and ensure it is not stale or moldy. Chopping the hay with scissors, or a food processor, and sprinkling, spraying, or briefly soaking the hay with water improves palatability. Soaking too long will leach out nutrients. Short hay segments are recommended to decrease colonic obstructions. Mixing greens in or under the hay also helps with acceptance. Pellets can be soaked in just enough water to be absorbed, until soft, and mixed into the greens. Be patient and persistent, tortoises and iguanas will switch over to commercial pellets, or hay (for tortoises), once hungry enough. Sulcata, as well as other giant tortoise species, grow so large and eat so much, that reliance on grass hays often becomes a practical necessity.

Consideration needs to be given to diluting commercial diets with natural foods, and creating an unbalanced ration. By definition complete diets are intended as sole diets, unless the feeding instructions specify they are formulated to be fed with natural foodstuffs. Some manufacturers have anticipated this, and Mazuri recommends feeding their pelleted tortoise diets with good quality grass hay, and, if desired, up to 5% fruit and 20% vegetables, by weight, of total diet.

Misconceptions: Be aware of several persistent widespread misconceptions. Members of the Brassica family (cabbages, kale, mustard greens, broccoli, cauliflower, Brussels sprouts, bok choy) are thought, but not proven, to cause thyroid problems (goiter) if fed exclusively long term, but are harmless in moderation as part of a balanced diet. Goiter appears exceedingly rare in tortoises, and goiter association with feeding Brassicas has never been validated in tortoises. Galapagos and Aldabra tortoises may have enlarged thymuses in their ventral neck, but these should not be confused with goiter. Massively enlarged thyroid glands are generally tumors, not goiter.

Many greens contain oxalates which can bind Ca in the intestinal tract and decrease calcium absorption. The adverse effects of oxalates must be considered in terms of oxalate:Ca molar ratio of food.[50] Spinach, beet greens, sorrels (*Oxalis* and *Rumex* spp.) and purslane (*Portulaca* spp.) have a oxalate:Ca ratio > 2, therefore they have no utilizable Ca, and possess excess oxalate, which can bind Ca in other foods eaten at the same time. They are not good sources of Ca despite their apparently high Ca levels.[50] Other foods, such as prickly pear cactus (*Opuntia ficus indica*), collards, Swiss chard, dandelions, kale, turnip greens, Brussel sprouts, and parsley, have a oxalate:Ca ratios < 2 indicating that they are good sources of Ca. Diets low in Ca and high in oxalates are not recommended, but occasional consumption of high oxalate foods, as part of a nutritious diet, does not pose any particular health problem.[50] Indeed, it is recommended to feed these greens, just not exclusively. Cattle and sheep, as well as humans, possess a bacteria, *Oxalobacter formigenes,* that can metabolize oxalic acid, perhaps herbivorous reptiles do as well.[50] Another unfounded concern with oxalates is that they can cause urocystoliths, probably because they may contribute to renoliths in humans. Reptiles form urate stones, not Ca oxalate stones. There are only two reports of Ca oxalate stones in tortoises, one in a Moorish tortoise (*Testudo graeca graeca*, the cystourolith was only 20 % calcium oxalate)[51] and one in a deceased wild desert tortoise with renoliths.[52] The latter scientific study concluded that oxalates in desert tortoises were an incidental finding and non-pathogenic.[52] In contrast, humans do get Ca oxalate stones, yet oxalate rich foods are still widely and safely consumed, and only restricted in people predisposed to Ca oxalate stone formation.

Plant poisoning is extremely rare in tortoises. Leopard tortoises (*Stigmochelys pardalis*) and tent tortoises (*Psammobates oculifer*) regularly eat the invasive African species, goat's head (*Tribulus terrestris*), which causes staggers (limb paresis) and death in livestock.[53,54] Leopard tortoises are known to consume more than a half dozen plant species suspected to contain toxic substances, including oxalates.[53] Most of the plants consumed by Russian tortoises, (*Agrionemys horsfeildi*), are notorious for their toxicity to herbivorous mammals.[55] More than 10% of plant species regularly consumed by Hermann's tortoises (*Testudo hermanni*) contained toxic alkaloids.[56] Tortoises will typically avoid poisonous plants, dilute their effects with other foods, or are resistant to their effects. Some exceptions include rhododendrons (*Rhododendron* spp., grayanotoxins cause flaccid paresis), oleanders, chinaberry trees, tree tobacco and toadstools.

Some important points for guidance are summarized here. All herbivores require UVB light. Regarding diet for both tortoises and iguanas, it is important to concentrate more on dietary variety than limiting foods. For tortoises, grasses, grass hays, native plants and good quality commercial tortoise diets should make up at least 80% of the diet. The remaining 20% can be dark leafy greens, Mulberry tree leaves, grape leaves, back yard weeds and non-starchy vegetables, as well as calcium rich flowers (roses, *Rosa* spp., nasturtiums, *Tropaeolum* spp., carnations, *Dianthus caryophyllus*, cactus flowers). For iguanas, produce should be no more than 50% of the diet, on an AF basis, with the remainder being a rationally formulated commercial diets.[2] Produce should be limited to dark leafy greens and non-starchy vegetables, no fruits and starchy vegetables should be fed.[42] Additional multivitamins are not needed for iguanas or tortoises, as long as they consume commercial foods formulated for them, but be sure the Ca levels are where they should be, tortoises may need additional Ca.

Omnivores

Omnivores average dietary composition consists of 15-40% protein, 5-40% fat, and 20-75% carbohydrates.[28] Many aquatic turtles, box turtles, wood turtles, forest tortoises, sea turtles and lizards consume both plant and animal material. Many omnivores consume more animals as juveniles, to fuel increased demands for protein and fat, than as adults.[29]

Box turtles: Box turtles are more carnivorous than initially thought and they eat whatever stimulates them and whatever they can catch, moving prey especially appeals to them.[57] Box turtles are opportunistic omnivores that consume beetles, insect larvae (caterpillars, grubs, maggots), cicadas, crickets, cockroaches, flies, grasshoppers, termites, millipedes, centipedes, land snails, slugs, earthworms, spiders, dragonfly larvae, sowbugs or pillbugs, crayfish, carrion, fish, frogs, tadpoles, toads, small mammals, birds, salamanders, lizards, snakes, smaller turtles, all types of carrion, as well as plant material, such as mushrooms (even poisonous ones), strawberries, raspberries, blackberries, blueberries, elderberries, mulberries, persimmons, sea grapes, tomatoes, apples, grasses, leaves, shoots, buds, roots, mosses and Opuntia fruits, pads and flowers, to name but a few.[57] Again fruits should be limited and good quality commercial diets designed for box turtle or aquatic turtles should make up a large portion of the diet.

Bearded dragons: One study of stomach contents of ten bearded dragons found a wide variety of arthropods (9 orders, 80% Isopterans [termites]), but a larger volume of plant matter. The diets were high in crude protein (41-50% DM), fatty acid content was 14-27% of the DM intake. Mealworms, most closely resembled the nutritional profile of termites, and the authors concluded a diet with a variety of insects and dark leafy greens would closely resemble the natural diet.[58] Another study showed adult bearded dragons consumed 90% plant material, where as juveniles consumed 50% plant material and 50% animal material.[59] Bearded dragons are also lizard eaters and cannibalistic if size disparity is present. Suitable vegetables include calcium-dusted dark leafy greens, carrots, squash, zucchini, peas and beans. Flowers such as roses, nasturtiums, carnations and hibiscus are also good. Dragons should be fed a wide variety of gut loaded and dusted insects. Commercial pelleted (not cubed) bearded dragon foods can also be fed. Baby mice can be offered several times per month. Even though bearded dragons enjoy fruits, they are not recommended. Overfeeding and obesity are major problems in captivity. Growing juveniles can be fed daily, while adults should be fed 2-3 times weekly. Growing dragons need more calcium than adults. Calcium is provided in calcium-rich insect gut-loading diets, dusted on insects or sprinkled on food. If multivitamins are not included in the gut loading, dusting preparation, or fortified foods, multivitamins can be given twice a month.

Aquatic turtles: Many large Emydid turtles including *Chrysemys, Graptemys, Pseudemys*, and *Trachemys* spp. undergo an ontogenetic shift in diet as they mature.[60] The red eared slider (*Trachemys scripta*) starts out predominantly carnivorous as juveniles, inhabiting shallower warmer waters that are rich in invertebrate and animal prey (the later are good sources of protein and Ca). As they get older, and larger, they cannot maneuver in shallow water and move into deeper cooler water (typically 5-6 °C lower) where prey is more dispersed, and so they shift to eating abundant plant material, with lower protein content, such as duckweed (*Spirodela polyrhiza*) and hydrilla (*Hydrilla verticillata*).[60]

Conclusions

Veterinarians have been trained in animal nutrition. The herp veterinarian must evaluate dietary history on a regular basis to keep their reptile and amphibian patients healthy. Information within this chapter covers the majority of reptiles and amphibians seen by veterinarians in private and zoological veterinary practice. With basic information and as vocal advocates for good nutrition, veterinarians can have a tremendous impact on the health of captive herps. Be proactive.

References

1. Lillywhite HB. Behavior and physiology: an ecological and evolutionary viewpoint on the energy and water relations of ectothermic amphibians and reptiles. de Andrade DV, Bevier CR, de Carvalho JE, eds. *Amphibian and Reptile Adaptations to the Environment*. Boca Raton, FL: CRC Press;2016;1-40.

2. Allen M, Oftedal O. Nutrition in captivity. In: Jacobson ER, ed. *Biology, Husbandry and Medicine of the Green Iguana*. Malabar, FL: Krieger Publishing;2003:47-74.

3. Donoghue S, Dzanis DA. 1995. Evaluating commercial diets. *Proc Annu Conf Assoc Reptilian Amphib Vet.* 74-79.

4. Delaney SJ, Fascetti AJ. 2012. Using pet food labels and product guides. In: Fascetti AJ, Delaney SJ, eds. *Applied Veterinary Clinical Nutrition.* Ames, IA: Wiley-Blackwell;69-74.

5. Webb GJW, Hollis GJ, Manolis SC. Feeding, growth, and food conversion rates of wild juvenile saltwater crocodiles (*Crocodylus porosus*). *J Herpetol.* 1991;25(4):462-473.

6. National Research Council. Vitamin A. In: *Vitamin Tolerance of Animals.* Washington, DC: National Academy Press;1987:3-10.

7. Ward AM, Dempsey JL. Nutrition. In: Lemm JM, Alberts AC, eds. *Cyclura: Natural History, Husbandry, and Conservation of West Indian Rock Iguanas*. London, UK: Academic Press;2012:129-142.

8. Vitt LJ, Caldwell JP. *Herpetology: An Introductory Biology of Amphibians and Reptiles.* 4th ed. St. Louis, MO: Elsevier;2014:(291):309-113.

9. Bernard JB, Allen ME. Feeding captive insectivorous animals: nutritional aspects of insects as food, fact sheet 003. *Nutrition Advisory Group Handbook*. 1997;1-7.

10. Barker D, Fitzpatrick MP, Dierenfeld ES. Nutrient composition of selected whole invertebrates. *Zoo Biol.* 1998;17(2):123-134.

11. Finke MD. Complete nutrient composition of commercially raised invertebrates used as food for insectivores. *Zoo Biol.* 2002;21(3):269-285.

12. Latney L, Clayton LA. Updates on amphibian nutrition and nutritive value of common feeder insects. *Vet Clin North Am Exot Anim Pract.* 2014;17:347-367.

13. Meiron OE, Bar-David E, Aflalo ED, et al. Solubility and bioavailability of stabilized amorphous calcium carbonate. *J Bone Miner Res.* 2011:25:364-372.

14. Allen M, Oftedal O. Dietary manipulation of the calcium content of feed crickets. *J Zoo Wildl Med.* 1989;20(1):26-33.

15. Zwart P, Rulkens RJ. Improving the calcium content of mealworms. *Int Zoo Yearb.* 1979;19(1):254-255.

16. Strzelewicz MA, Ullrey DE, Schafer ST, et al. Feeding insectivores: increasing the calcium content of wax moth (*Galleria mellonella*) larvae. *J Zoo Anim Med.* 1985;16(1):25-27.

17. Klasing KC, Thacker P, Lopez MA. et al. Increasing the calcium content of mealworms (*Tenebrio molitor*) to improve their nutritional value for bone mineralization for growing chicks. *J Zoo Wildl Med.* 2000;31(4): 512-517.

18. Coslik AH, Ward AM, McClements RD. Gut loading as a method to effectively supplement crickets with calcium and vitamin A. *Proc Conf Zoo Wildl Nutr.* 2009;163-171.

19. Finke M, Dunham S, Kwabi C. Evaluation of four dry commercial gut loading products for improving the calcium content of crickets, *Acheta domesticus*. *J Herpetol Med Surg.* 2005;15(1):7-12.

20. Finke M, Dunham S, Cole J. Evaluation of various calcium-fortified high moisture commercial products for improving the calcium content of crickets, *Acheta domesticus*. *J Herpetol Med Surg.* 2004;14(2):17-20.

21. Granados D, Schmidt D, Schlegel M. Calcium gut loading of mealworms (*Tenebrio molitor*) and the benefit of gel water over apple slices for maximum calcium gut loading. *Proc Conf Zoo Wildl Nutr.* 2007;205-208.

22. Attard L. Lydia Attard. *The development and evaluation of a gut loading diet for feeder crickets formulated to provide a balanced nutrient source for insectivorous amphibians and reptiles.* Master's thesis. University of Guelph, Guelph, Ontario, Canada.

23. Livingston S, Lavin SR, Sullivan K, et al. Challenges with effective nutrient supplementation for amphibians: a review of cricket studies. *Zoo Biol.* 2014;33:565-576.

24. Li H, Vaughan MJ, Browne RK. A complex enrichment diet improves growth and health in the endangered Wyoming toad (*Bufo baxteri*). *Zoo Biol.* 2009;28(3):197-213.

25. Michaels CJ, Antwis RE, Preziosi RF. Manipulation of the calcium content of insectivore diets through supplementary dusting. *J Zoo Aquar Res.* 2014;2:77-81.

26. Ferrie GM, Alford VC, Atkinson J, et al. Nutrition and health in amphibian husbandry. *Zoo Biol.* 2014;33(6):485-501.

27. Knight M, Glor R, Smedley, S, et al. Firefly toxicosis in lizards. *J Chem Ecol.* 1999;25(9):1981-1986.

28. Calvert, I. Nutrition. In: Girling SJ, Raiti P eds. *BSAVA Manual of Reptiles.* 2nd ed. Gloucester, UK: BSAVA Publications;2004:18-39.

29. Donoghue S. Nutrition. In: Mader D, ed. *Reptile Medicine and Surgery.* St Louis, MO: Elsevier Saunders;2006:251-298.

30. Allen M, Oftedal O. The nutrition of carnivorous reptiles. In: Murphy JB, Adler K, Collins JT, eds. *Captive Management and Conservation of Amphibians and Reptiles*. Washington, DC: Society for the Study of Amphibians Reptiles;1994:71-82.

31. Dierenfeld ES, Alcorn HL, Jacobsen KL. Nutrient Composition of Whole Vertebrate Prey (Excluding Fish) Fed in Zoos. Beltsville, MD: National Agricultural Library;2002:1-20.

32. American Veterinary Medical Association. AVMA Guidelines for the Euthanasia of Animals. 2013 ed. Schaumburg, IL: AVMA;2013:48-50.

33. Veer V, Chiszar D, Smith H. Preference for formerly frozen mice over freshly euthanized mice by captive snakes of seven species from families Boidae and Colubridae. *Vivarium.* 1996;7(6):18.

34. Bernard JB, Allen ME. Feeding captive piscivorous animals: nutritional aspects of fish as food, fact sheet 5. *Nutrition Advisory Group Handbook*. 1997:1-7.

35. Britton A. Diet, feeding and nutrition. In: *Crocodilian Biology and Captive Management*, St. Augustine, FL: AAZA Professional Development Program;2010:1-33.

36. Baer DJ. Nutrition in the wild. In: Jacobson ER ed. *Biology, Husbandry and Medicine of the Green Iguana.* Malabar, FL: Krieger Pubishing;2003:39-46.

37. Van Marken Lichtenbelt WD. Digestion in an ectothermic herbivore, the green iguana (*Iguana iguana*): effect of food composition and body temperature. *Physiol Zool.* 1992;65(3):649-673.

38. Rand AS. Reptilian arboreal folivores. In: Montgomery GG, ed. *The Ecology of Arboreal Folivores.* Washington, DC: Smithsonian Institution Press;1978:115-122.

39. Esque TC, Drake KK, Nussear NE. Water and food acquisition and their consequences on life history and metabolism of North American tortoises. In: Rostal DC, McCoy ED, Mushinsky, HR, eds. *Biology and Conservation of North American Tortoises.* Baltimore, MD: Johns Hopkins University Press;2014:85-95.

40. Oftedal OT. Nutritional ecology of the desert tortoise in the Mojave and Sonoran deserts. In: Van Devender TR, ed. *The Sonoran Desert Tortoise Natural History, Biology and Conservation*. Tucson, AZ: University of Arizona Press;2002:194-241.

41. Jarchow J. Veterinary management of the desert tortoise, *Gopherus agassizii* at the Arizona-Sonora Desert Museum: a rational approach to diet. *Proc Annu Conf Symp Desert Tortoise Counc.* 1984;83-94.

42. Parsons JL. Feeding captive desert tortoises: appropriate diets for arid zone herbivorous tortoises. *Proc Annu Conf Symp Desert Tortoise Counc.* 2012;32-33.

43. Moberly WR. The metabolic responses of the common iguana, *Iguana iguana*, to activity under restraint. *Comp Biochem Physiol.* 1968;27:1-20.

44. Troyer KA. Diet selection and digestion in *Iguana iguana*: the importance of age and nutrient requirements. *Oecologia.* 1984;61:201-207.

45. Rosskopf WJ, Howard E, Gendron AP, et al. Mortality studies on *Gopherus agassizi* and *Gopherus berlandieri* tortoises. *Proc Annu Conf Symp Desert Tortoise Counc.* 1981;108-112.

46. Burge BL, Bradley WB. Population density, structure, and feeding habits of the desert tortoise, *Gopherus agassizii*, in a low desert study area in southern Nevada. *Proc Annu Conf Symp Desert Tortoise Counc.* 1976:51-74.

47. Hansen RM, Johnson MK, Van Devender TR. Foods of the desert tortoise, *Gopherus agassizii*, in Arizona and Utah. *Herpetologica.* 1976;32:247-251.

48. Coombs EM. Status of the desert tortoise, *Gopherus agassizii*, in the state of Utah. *Proc Annu Conf Symp Desert Tortoise Counc.* 1977; 95-101.

49. Ullrey, DE. Hay quality evaluation, fact sheet 001. *Nutrition Advisory Group Handbook.* 1997;1-10.

50. Noonan SC, Savage GP. Oxalate content of foods and its effect on humans. *Asia Pacific J Clin Nutr.* 1999;8(1):64-74.

51. Mathes K, Gunther P, Kowaleski N, Fehr M. Urinary calculus in the pelvic region of a Moorish tortoise. *Tierarztliche Praxis.* 2009;37(6):427-432.

52. Jacobson ER, Berry KH, Stacy B, et al. Oxalosis in wild desert tortoises, *Gopherus agassizii*. *J Wildl Dis.* 2009;45(4):982-988.

53. Kabigumila J. Sighting frequency and food habits of the leopard tortoise, *Geochelone pardalis*, in northern Tanzania. *Afr J Ecol.* 2001;39:276-285.

54. Rall M, Fairall N. Diet and food preferences of two South African tortoises, *Geochelone pardalis* and *Psammobates oculifer*. *S Afr J Wildl Res.* 1993;23:63-70.

55. Lagarde F, Bonnet X, Corbin J, et al. Foraging behavior and diet of an ectothermic herbivore: *Testudo horsfieldi*. *Ecography.* 2003;26(2):236-242.

56. Meek R. Aspects of the ecology of *Testudo hermanni* in southern Yugoslavia. *Br J Herpetol.* 1985;6: 437-445.

57. Dodd K. *North American Box Turtles: A Natural History*. Norman, OK: University of Oklahoma Press;2001.

58. Oonincx DG, van Leeuwen JP, Hendriks WH, van der Poel AF. The diet of free-roaming Australian central bearded dragons (*Pogona vitticeps*). *Zoo Biol.* 2015;34(3):271-277.

59. MacMillen RE, Augee ML, Ellis BA. Thermal ecology and diet of some xerophilous lizards from western New South Wales (Australia). *J Arid Environ.* 1989;16:193-201.

60. Avery HW, Spotila JR, Congdon JD, et al. Roles of diet protein and temperature in the growth and nutritional energetics of juvenile slider turtles, *Trachemys scripta*. *Physiol Zool.* 1993;66(6):902-905.

ARAV
ASSOCIATION OF REPTILIAN AND AMPHIBIAN VETERINARIANS

Section 22

ARAV Nutrition Bonus Session

Susan Horton, DVM and
Nicola Di Girolamo, DMV, MSc (EBHC), GPCert(ExAP), PhD, DipECZM (Herpetology)
Session Chairs

Evaluation of Carotenoid Assimilation in Leopard Geckos (*Eublepharis macularius*)

Ophélie Cojean, DVM, IPSAV (Zoological Medicine),
Stéphane Lair, DMV, DES, DVSc, Dipl ACZM,
Claire Vergneau-Grosset, DVM, IPSAV (Zoological Medicine), Dipl ACZM

Session #2012

Affiliation: From Clinique Vétérinaire Alliance, 8 Boulevard Godard, 33300 Bordeaux, France (Cojean), Faculté de médecine vétérinaire, Département de sciences cliniques, Université de Montréal, Canada and the Canadian Cooperative Wildlife Health Centre, Quebec, Canada (Lair), Université de Montréal, 3200 rue Sicotte, Saint-Hyacinthe, Quebec, J2S2M2, Canada Vergneau-Grosset).

Although leopard geckos (*Eublepharis macularius*) are commonly kept under human care, their vitamin requirements are largely unknown. Many invertebrate preys display a low concentration of vitamin A; thus gut-loading insects with vitamin A or carotenoids is a common practice. The objective of this prospective experimental study was to investigate whether leopard geckos can absorb and assimilate oral carotenoids.

Ten clinically healthy female leopard geckos were randomly divided in 2 groups with various supplementations: a group receiving vitamin A supplementation and a group receiving carotenoids. Insects were gut-loaded continuously with a supplement containing vitamin A or carotenoids, depending on the group. Oral supplementation with cod liver oil or carrot juice was administered weekly to each lizard from group vitamin A and group carotenoid respectively.

After 10 weeks of supplementation, surgical hepatic biopsies were obtained in 3 geckos of each group while the 2 remaining geckos were euthanatized to undergo complete necropsy. Hepatic retinol concentration was determined for each lizard ($n = 10$) by ultra performance liquid chromatography. Histopathology revealed hepatocellular vacuolization and vitellogenic follicles in 5 females. Epithelial squamous metaplasia was not observed in any of the geckos. Hepatic retinol concentration was significantly higher in the group carotenoid than in the group vitamin A ($P = 0.03$).

Contrary to the herbivorous green iguana (*Iguana iguana*), our results suggest that leopard geckos could absorb and assimilate oral carotenoids.

Acknowledgments: This project was funded by the Association of Reptilian and Amphibian Veterinarians 2015-2016 grant.

Section 23

ARAV Reproduction

Susan Horton, DVM and
Nicola Di Girolamo, DMV, MSc (EBHC), GPCert(ExAP), PhD, DipECZM (Herpetology)
Session Chairs

Ultrasonographic Anatomy of Reproductive Female Leopard Geckos (*Eublepharis macularius*)

Ophélie Cojean, DVM, IPSAV (Zoological Medicine),
Claire Vergneau-Grosset, DVM, IPSAV (Zoological Medicine), Dipl ACZM,
Isabelle Masseau, DVM, PhD, MSc, Dipl ACVR

Session #2013

Affiliation: From Clinique Vétérinaire Alliance, 8 Boulevard Godard, 33300 Bordeaux, France (Cojean), Université de Montréal, 3200 rue Sicotte, Saint-Hyacinthe, Quebec, J2S2M2, Canada Vergneau-Grosset, Masseau).

Captive leopard geckos (*Eublepharis macularius*) often present to the exotic clinic for gastrointestinal impactions, follicular stasis, or dystocia. To our knowledge, normal ultrasonographic anatomy of these lizards has not been described.

Ultrasound examinations of the head, coelomic cavity and tail were performed in eleven healthy female leopard geckos. The objectives of this study were to describe the normal sonographic anatomy of the head, coelomic cavity and tail as well as to assess a potential relationship between body weight and measures of the proximal aspect of the tail and of the intracoelomic fat bodies. Our hypothesis was that the tail circumference would be a good indicator of body condition score in leopard geckos.

Linear array 13-18MHz transducer was used. Leopard geckos were sedated and restrained in dorsal recumbency for coelomic structures examination, and in ventral recumbency for brain and tail examinations. Sagittal and transverse images were obtained. A descriptive analysis was conducted.

The ventral surface of the lungs, liver, gallbladder, caudal vena cava, portal vein, ventral abdominal vein, aorta, ovarian follicles, fat bodies, tail and brain were visualized in 10 of 11 individuals. In 1 individual, molt precluded ultrasonographic examination. The heart, kidneys, urinary bladder, spleen and pancreas were not visualized. Digestive tract was observed in 10 individuals but was too small to be measured. Finally, there was a significant positive association between body weight and tail circumference in leopard geckos. These results could be used as a reference for the normal ultrasonographic anatomy of leopard geckos.

Health Effects of Repeated Electroejaculations on Veiled Chameleons (*Chamaeleo calyptratus*)

Sean M. Perry, DVM,
Michael Lierz, DZooMed, DipECZM (Wildlife Population Health), DipECPVS,
Mark A. Mitchell, DVM, MS, PhD, DipECZM (Herpetology)

Session #2076

Affiliation: From Louisiana State University, Veterinary Clinical Sciences, Skip Bertman Drive, Baton Rouge, LA 70810, USA (Perry, Mitchell), Justus-Liebig-University of Giessen, Frankfurter Str 91-93, Giessen, Germany (Lierz).

Abstract: Semen collection in most lizards and chelonians requires the use of electroejaculation. Successful semen collection using electroejaculation has been reported in the green iguana, leopard gecko, and leopard tortoise. Despite studies using electroejaculation to collect semen, no studies have evaluated whether repeated electroejaculation has any negative health effects on reptiles. A prospective longitudinal study was performed using 12 veiled chameleons (*Chamaeleo calyptratus*) to evaluate the health effects of repeated electroejaculations. Physical examinations, complete blood counts, and plasma chemistries were performed prior to and at the study's completion. Electroejaculations were conducted once weekly over 4 weeks. Animals were anesthetized for each procedure with alfaxalone 10 mg/kg IV. Animals were electroejaculated by using 15 cloacal intromissions at 0.1 mA. Animals were then given a rest period of 3 minutes between each series of electroejaculations. Animals underwent a maximum of 3 series of electroejaculations. Once an ejaculate was collected, electroejaculation was discontinued. Photographic documentation of the external structures of the vent were collected prior to and after each anesthesia event. Cloacoscopy was performed before and after each anesthetic event to evaluate the internal cloacal structures for any changes post-electroejaculation. External vent structures did not change over the course of the study. Mild inflammation was observed on cloacoscopy after electroejaculation; however, no chronic gross changes were observed in the cloacal tissues. All animals survived the study in good health, demonstrating that repeated weekly electroejaculations can be performed successfully in the veiled chameleons.

Determining Semen Parameters in Panther Chameleons (*Furcifer pardalis*): Developing a Model for Male Assisted Reproduction in Chameleons

Sean M. Perry, DVM,
Michael Lierz, DZooMed, DipECZM (Wildlife Population Health), DipECPVS,
Mark A. Mitchell, DVM, MS, PhD, DipECZM (Herpetology)

Session #2077

Affiliation: From Louisiana State University, Veterinary Clinical Sciences, Skip Bertman Drive, Baton Rouge, LA 70810, USA (Perry, Mitchell), Justus-Liebig-University of Giessen, Frankfurter Str 91-93, Giessen, Germany (Lierz).

Abstract: Assisted reproductive technologies have become an integral part of in situ and ex situ conservation programs for threatened and endangered species; however, very few studies have investigated these technologies for reptiles. Chameleons represent a family of reptiles where more than half of all documented species are threatened. The purpose of this study was to begin developing reproductive technologies for male chameleons using panther chameleons (*Furcifer pardalis*) as a model. Animals were anesthetized for each electroejaculation with alfaxalone 10 mg/kg IV. Animals were electroejaculated by using 15 cloacal intromissions at 0.1 mA. Animals were then given a rest period of 3 minutes between each series of electroejaculations. Animals underwent a maximum of 3 series of electroejaculations. Semen was collected with a 1-ml syringe and volume and color were recorded. Once an ejaculate was collected, electroejaculation was discontinued. A computer-assisted sperm analysis program was used to evaluate concentration, motility, rate of movement, plasma membrane integrity and acrosome integrity for the panther chameleons. Semen collection was performed bi-monthly. Sperm concentration was estimated using 2 different methods; a manual hand count with a hemocytometer and with an automated counter (Nucleocounter SP- 100). Morphology of spermatozoa was evaluated by observing 100 spermatozoa under oil immersion microscopy. The percentage of each type of abnormal spermatozoa were recorded. Semen samples were collected from the panther chameleons on a consistent basis.

Section 24

ARAV Infectious Diseases

Rachel Marschang, PD, Dr med vet, DipECZM (Herpetology), FTA and
Kim Wojick, DVM
Session Chairs

Salmonella enterica subspecies *arizonae*- and *salamae*-Associated Osteomyelitis Outbreak in a Herpetological Collection

Tom Hellebuyck, Dr med vet, DipECZM (Herpetology),
Francis Vercammen, Dr med vet,
Sylvie Pirson, med vet,
Marc Verlinden, Dr med vet,
Frank Pasmans, Dr med vet, DipECZM (Herpetology)

Session #2072

Affiliation: From the Department of Pathology, Bacteriology and Avian Diseases, Faculty of Veterinary Medicine, Ghent University, B-9820 Merelbeke, Belgium, Centre for Research and Conservation, Royal Zoological Society of Antwerp, Koningin Astridplein 26, B-2018 Antwerp, Belgium.

Three corn snakes (*Pantherophis guttatus*) belonging to a public herpetological collection were presented with chronic weight loss, distention of the caudal coelomic cavity and muscle flaccidness. Clinical examination revealed the presence of firm, subcutaneous nodules located at the dorsal midline of the lumbar region. Whole-body lateral and ventrodorsal radiographs revealed osteolysis of the bodies of the several lumbar vertebrae as well as extensive new bone production associated with multiple sites in the vertebral column in all snakes. *Salmonella enterica* subspecies *arizonae* serotype 44:Z4Z23 and *S. enterica* subspecies *salamae* serotype 18:Z4Z23 were isolated following microbiologic examination of fine needle aspirates of the affected vertebrae in 2 and 1 of the snakes respectively. During necropsy, multiple, yellowish white, macroscopic nodules were observed at the ventral side of multiple vertebrae. Severe stenosis of the lumbar spine was noted on cross section of the infected vertebrae. In addition, granulomatous orchitis and hepatitis and colonic dilatation with stasis of intestinal content were noted in 2 snakes. *S. enterica* subspecies *arizonae* serotype 44:Z4Z23 was isolated from the testicles and the liver. *Salmonella* spp. could not be isolated from faecal samples of the affected snakes and their healthy cage mates. Although histologic examination did not reveal hypertrophia of the gastric wall, *Cryptosporidium serpentis* infection was confirmed in all affected snakes. Three months later, vertebral osteomyelitis associated with *S. enterica* subspecies *arizonae* serotype 44:Z4Z23 infection was demonstrated in 2 corn snakes of a separately housed group of 6 corn snakes in the same collection showing identical clinical signs as described above.

Isolation of the Highly Pathogenic and Zoonotic Agent *Burkholderia pseudomallei* from a Pet Green Iguana (*Iguana iguana*) in Belgium

Tom Hellebuyck, Dr med vet, DipECZM (Herpetology),
Pierre Wattiau, Dr,
Filip Boyen, Dr med vet,
Ilse Moeremans, med vet,
Frank.Pasmans, Prof Dr med vet, DipECZM (Herpetology),
Freddy Haesebrouck, Prof Dr med vet

Session #2073

Affiliation: From the Department of Pathology, Bacteriology and Avian Diseases, Faculty of Veterinary Medicine, Ghent University, B-9820 Merelbeke, Belgium, Department of Bacteriology and Immunology, Veterinary and Agrochemical Research Centre, Groeselenberg 99, B-1180 Brussels, Belgium.

A pure and abundant culture of gram-negative rods that could not be identified using conventional phenotypic systems was obtained from liver abscesses in a 5-year-old pet green iguana (*Iguana iguana*) following necropsy. The isolate was identified as *Burkholderia pseudomallei* by multiple-locus sequence typing-derived polymerase chain reaction (PCR) and real-time PCR.[1] The isolation of the highly pathogenic and zoonotic agent *B. pseudomallei* has been previously reported from a pet green iguana in the Czech republic with dermal abscesses and 2 pet iguanas in the United States with oral, dermal and liver abscesses.[2,3] *B. pseudomallei* is the causative agent of melioidosis. Melioidosis is endemic in countries in Southeast Asia and Northern Australia, and an emerging infectious disease in various tropical developing countries.[4] As previously reported, infected humans and import of infected animals can introduce melioidosis to non-endemic areas as the bacterium may be released and persist in the environment.[4,5] Infection in humans commonly occurs through cutaneous inoculation and ingestion with soil or water or inhalation of aerosolized bacteria.[4] The present case highlights the risk of imported green iguanas acting as a reservoir for the introduction of this high-threat zoonotic pathogen into non-endemic regions. Awareness of this potential risk should be raised in veterinarians dealing with pet iguanas originating from endemic areas and physicians that are consulted by iguana pet owners with symptoms compatible with melioidosis.

References

1. Wattiau P, Van Hessche M, Neubauer H, et al. Identification of *Burkholderia pseudomallei* and related bacteria by multiple-locus sequence typing-derived PCR and real-time PCR. *J Clin Microbiol.* 2007;45(3): 1045-1048.

2. Elschner MC, Hnizdo J, Stamm I, et al. Isolation of the highly pathogenic and zoonotic agent *Burkholderia pseudomallei* from a pet green iguana in Prague, Czech Republic. *BMC Vet Res.* 2014;10(283):1-5.

3. Zehnder AM, Hawkins M, Koski MA, et al. *Burkholderia pseudomallei* isolates in 2 pet iguanas, California, USA. *Emerg Infect Dis.* 2014;20(2):304-306.

4. Limmethurotsakul D, Golding N, Dance DAB, et al. Predicted global distribution of *Burkholderia pseudomallei* and burden of melioidosis. *Nat Microbiol.* 2016;1(1):15008.

5. Galimand M, Dodin A. Repartition de *Pseudomonas pseudomallei* en France et dans le monde la melioidose. *Bull Soc Vet Prat de France.* 1982;66(3):651-657.

Experimental Inoculation of Boa Constrictors (*Boa constrictor*) with Reptarenaviruses

David Sanchez-Migallon Guzman, LV, MS, DipECZM (Avian), DipECZM (Small Mammal), Dipl ACZM,
Mark D. Stenglein, PhD,
Michelle Hawkins, VMD, DABVP (Avian Practice),
Valentina Garcia,
Marylee L. Layton,
Laura L. Hoon-Hanks,
Scott M. Boback,
M. Kevin Keel, DVM, PhD, DACVP,
Tracy Drazenovich, DVM,
Joseph L. DeRisi, PhD

Session #2116

Affiliation: From the Department of Veterinary Medicine and Epidemiology, School of Veterinary Medicine, University of California, Davis, CA 95616, USA (Guzman, Hawkins, Drazenovich), Department of Microbiology Immunology and Pathology, College of Veterinary Medicine and Biomedical Sciences, Colorado State University (Stenglein, Layton, Hoon-Hanks), Department of Biology, Dickinson College, Carlisle, PA (Boback), Department of Pathology, Microbiology and Immunology, School of Veterinary Medicine, University of California, Davis, CA 95616, USA (Keel), Department of Biochemistry and Biophysics, School of Veterinary Medicine, University of California San Francisco, San Francisco, CA 94158, USA (Garcia, DeRisi).

Inclusion body disease (IBD) is a fatal disease of snakes. IBD is most commonly diagnosed in boas and pythons, but has been reported in several other species, and disease manifestation is variable in different species.[1] Reptarenaviruses have been associated with IBD in boid snakes,[2,3] and have been shown to cause severe neurologic signs and lesions (moderate encephalomyelitis, meningitis and ganglioneuritis) in ball pythons inoculated experimentally in a prior study.[4] For this study, 3 boa constrictors that tested negative on RNA qRT-PCR for reptarenavirus from blood and liver biopsies were used. Two boa constrictors were inoculated with the reptarenavirus Golden Gate virus (GoGV) while 1 additional boa constrictor served as a control. The virus used to inoculate the 2 snakes were isolated from previous boa constrictors with inclusion body disease. The infected snakes were monitored for 2 years, without showing any clinical signs. Repeated antemortem liver and lung biopsies and postmortem examination revealed characteristic inclusion bodies in multiple tissues including the central nervous system. The presence of viral RNA for reptarenavirus was confirmed and quantified by qRT-PCR in the infected snakes and immunofluorescence was performed in multiple tissues demonstrating the presence of the virus. The control snakes remained without clinical signs during the same period and no significant lesions were found in histopathology. There was no reptarenavirus RNA found in the control snake and immunofluorescence was negative. The results of this study establish a causative relationship between reptarenavirus and inclusion bodies in boa constrictors.

Acknowledgments: This study was supported by the Reptile Research Fund at the University of California, Davis and the Howard Hughes Medical Institute.

References

1. Chang L-W, Jacobson ER. Inclusion body disease, a worldwide infectious disease of boid snakes: a review. *J Exot Pet Med.* 2010;19(3):216-225.

2. Stenglein MD, Sanders C, Kistler AL, et al. Identification, characterization, and in vitro culture of highly divergent arenaviruses from boa constrictors and annulated tree boas: candidate etiological agents for snake inclusion body disease. *mBio.* 2012;3:e00180-00112.

3. Stenglein MD, Jacobson ER, Chang LW, et al. Widespread recombination, reassortment, and transmission of unbalanced compound viral genotypes in natural arenavirus infections. *PLoS Pathog* 2015;11:e1004900.

4. Sanchez-Migallon Guzman D, Stenglein MD, Hawkins M, et al. Experimental inoculation of ball pythons (*Python regius*) with snake arenaviruses. *Proc Annu Conf Assoc Reptilian Amphib Vet.* 2016;693-694.

Hematologic Changes in Cottonmouth Snakes (*Agkistrodon piscivorus*) Infected With and Without Snake Fungal Disease

Jessica Huntington, BS,
Sarah Baker, PhD,
Ethan Kessler, BS,
Matthew C. Allender DVM, MS, PhD, Dipl ACZM

Session #2089

Affiliation: From the Department of Comparative Biosciences, Wildlife Epidemiology Lab, College of Veterinary Medicine, University of Illinois, 2001 S. Lincoln Ave, Urbana, IL 61802 USA (Huntington and Allender), Illinois Natural History Survey, Prairie Research Institute, University of Illinois, 1816 S. Oak St, Champaign, IL 61820 USA (Baker, Kessler, Allender).

Five cottonmouth snakes (*Agkistrodon piscivorus*) were experimentally inoculated with *Ophidiomyces ophiodiicola* and complete blood counts were performed serially (bi-weekly) for 90 days. Nasolabial skin swabs were paired for qPCR analysis. Packed cell volume and total solids analysis were performed on all samples. Blood smears were made and stained with diff-quick for a 100 white blood cell differential count and a quantitative analysis of hemoparasites. On a repeated measures ANOVA, packed cell volume and total solids were significantly lower post inoculation ($P < 0.0001$ and $P = 0.0121$ respectively) when compared to pre-inoculation values. There were no significant changes seen in pre and post inoculation states with the WBC count or with the quantity of hemoparasites. The findings of this study conclude that there is a mild hematologic effect in individuals infected with *O. ophiodiicola*. Future studies are needed to further understand the hematologic changes seen in snakes that are infected with *O. ophiodiicola* and evaluate other aspects of the host response. Utilizing other diagnostic assays may further elucidate snake fungal disease to aid in clinical and conservation plans.

Section 25

ARAV Clinical Cases and Reviews

Tom Torres, DVM and
Erica Giles, DVM
Session Chairs

Successful Treatment of Firefly Intoxication in a Bearded Dragon (*Pogona vitticeps*)

David Hannon, DVM, DABVP (Avian Practice)

Session #2015

Affiliation: From Avian and Exotic Animal Veterinary Services, Memphis Veterinary Specialists, Memphis, TN 38018, USA.

An adult bearded dragon (*Pogona vitticeps*) presented moribund several hours after being fed a firefly. The owner had noted that the lizard had vomited in the enclosure. Initial labwork showed marked leukocytosis, hyperglycemia, hyperkalemia, and hypophosphatemia. Treatment was initiated with aggressive intraosseous fluid therapy, antiemetics, and repeated administration of activated charcoal. The lizard responded favorably after 4 days in the hospital and was eating well prior to discharge. The hyperglycemia had resolved when rechecked 2 weeks later. The successful treatment outcome may have been associated with regurgitation by the lizard after firefly ingestion, or because aggressive treatment was initiated very quickly after ingestion.

Preliminary Results of the Surveillance of Physiologic Values of Sea Snakes in the Gulf of Oman

Csaba Géczy, DVM,
Balázs Buzás,
Eszter Gulyás, MSc,
Ana Perez de Vargas, DVM, MS,
János Gál, DVM, PhD, DipECZM (Herpetology)

Session #2053

Affliation: From the Department of Exotic Animal and Wildlife Medicine, University of Veterinary Science, István u. 2., H-1078, Budapest, Hungary (Gál), Private, Al Ain, Abu Dhabi, United Arab Emirates (Géczy, Perez de Vargas), Al Mayya Sanctuary, Fujairah, United Arab Emirates (Buzás, Gulyás).

Marine snakes are among the lesser studied reptiles, with majority of the research emphasizing on the ecology and less emphasize is placed on clinically relevant information. Goal of the surveillance was to gather data on basic physiologic values of sea snake species occurring in the area, including hematology, clinical chemistry, and bacterial culturing. The study area was the Gulf of Oman, in Fujairah, United Arab Emirates. Live snakes were monthly collected by scooping with a net from boat and placed in individual boxes. Snakes were transported to land for further examination and sample collection. Physical examination was performed on each snake prior sampling, sex was determined by cloacal probing, and basic morphometrics were recorded (weight, total length). Blood and fecal samples were collected within 12 hours of capture and kept refrigerated at 4 °C. Blood samples were collected from the ventral coccygeal vein, placed in lithium heparin containing tubes, and were sent to laboratory for analysis within 24 hours. Hematologic examination was performed by using standard methods, described in the literature, while PCV and Hb was determined by a commercial analyzer (Hemo Vet, EKF-diagnostic GmBH, Ebendorfer Chaussee 3, Technologiepark Ostfalen, D-39179 Barleben/Magdeburg, Germany). Clinical chemistry was performed by a commercial analyzer (Miura 500, ISE Srl, Via Luigi Einaudi, 7, 00012 Guidonia [RM], Italy). Fecal samples were processed for bacterial culturing within 24 hours after sample collection. Samples were incubated at 26 °C for a minimum of 3 days on various agar plates (blood agar, Hektoen, xylose lysine desoxycholate).

Of the 10 sea snake species naturally occurring in the region, 3 were regularly encountered: ornate sea snake (*Hydrophis ornatus*), Arabian-gulf sea snake (*H. lapemoides*), yellow-bellied sea snake (*H. platurus*). The preliminary results of hematology and chemistry showed no major difference compared to the previously published data of other terrestrial or marine species. Bacterial culturing yielded similar results than those of other reptile species. Commonly isolated bacteria included *Pseudomonas aeruginosa*, *Enterobacter cloaceae*, *Salmonella* sp., *Shigella* sp., and *Citrobacter freundii*.

Cloacal Scent Gland Pathology and Surgical Excision in Snakes: Case Series

Émilie L. Couture, DMV, IPSAV,
Nancy P. Lung, VMD, MS,
Stéphane Lair, DMV, DES, DVSc, Dipl ACZM,
Shannon T. Ferrell, DVM, DABVP (Avian Practice), Dipl ACZM

Session #2083

Affiliation: From the Faculté de Médecine Vétérinaire, Université de Montréal, 3200 rue Sicotte, Saint-Hyacinthe, QC J2S 2M2 Canada (Couture, Lair), Fort Worth Zoo, Fort Worth, TX 76110, USA (Lung), and Zoo de Granby, 525 rue Saint-Hubert, Granby, QC J2G 5P3 Canada (Ferrell).

Members of the suborder Serpentes possess paired scent glands located caudally to the vent. In this case series, 4 colubrids consisting of 1 Taiwan beauty snake (*Orthriophis taeniura freesei*), 1 black pine snake (*Pituophis melanoleucus lodingi*), 1 bullsnake (*Pituophis catenifer sayi*), and 1 central plains milksnake (*Lampropeltis triangulum gentilis*), 1 Trans-Pecos copperhead (*Agkistrodon contortrix pictigaster*), and 1 ball python (*Python regius*) presented with a cloacal scent gland impaction. The colubrids were greater than 20-years-old, and most (3/4) were females. Infection (3/4), mineralization (1/4), or neoplasia (1/4, fibrosarcoma) of the gland were associated with the impactions. When attempted (1/4), medical treatment was unsuccessful. Both glands were initially or eventually affected (4/4). The python was a 10-year-old male, and the viper was of undetermined age and male. In these latter snakes, gland impaction was unilateral. An underlying bacterial adenitis was diagnosed in the python, while the cause of impaction was not determined in the viper. Overall, complete surgical excision was done in 9/10 affected glands. Post-operative infection of the surgery site was encountered in the python and the Taiwan beauty snake (2/6). Colubrids died 6 weeks to 5 years post-operatively of causes unrelated to the surgery. No complications occurred in a 13-year follow-up of the viper, and the python has been doing well for 1.5 years. In conclusion, impaction of cloacal scent glands in snakes may be treated with surgical excision. Associated bacterial infections frequently need to be addressed, and any potential concurrent pathologic processes should be investigated. Based on these few cases and reported gland morphology, an association with age and sex is suspected in colubrids.

Acknowledgments: The authors would like to thank the Fort Worth Zoo, Zoo de Granby, and l'Aquarium du Québec for their assistance in the care of these snakes.

Cardiomyopathy and Aortic Aneurysm in a Bearded Dragon (*Pogona vitticeps*)

Anne Wyer, DVM,
Sue Chen, DVM, DABVP (Avian Practice),
Michael Garner, DVM, DACVP,
Michelle Fabiani, DVM, Dipl ACVR,
Natalie Antinoff, DVM, DABVP (Avian Practice)

Session #2086

Affiliation: From Veterinary Specialists and Emergency Services, 825 White Spruce Blvd, Rochester, NY 14623, USA (Wyer), Gulf Coast Veterinary Specialists Avian & Exotics, 1111 West Loop South, Houston, TX 77027, USA (Chen, Antinoff, Fabiani), Northwest ZooPath, 654 West Main, Monroe, WA 98272, USA (Garner).

A 4-year-old male bearded dragon was presented for a several week history of lethargy, anorexia, gular edema and bilateral periocular swelling that improved with heat support. Radiographs demonstrated a soft tissue mass within the cranial thorax. Echocardiographic measurements of the ventricle and left atrium were above the established reference ranges confirming that the radiographic findings were consistent with cardiomegaly. Pericardial effusion was appreciated. Clinical improvement was observed with the use of furosemide, enalapril, pimobendan and ceftazidime as well as a therapeutic pericardiocentesis. A large aortic aneurysm became apparent on both oral examination and repeated echocardiogram one month after initial examination. No additional treatment for the aneurysm was initiated, cardiac medications were continued. Rupture of the aneurysm lead to the decline and eventual euthanasia of the patient three months after initial presentation. Standard cardiac medications including furosemide, enalapril and pimobendan had a therapeutic effect in the treatment of presumed heart disease and aortic aneurysm in a bearded dragon.

ASSOCIATION OF REPTILIAN AND AMPHIBIAN VETERINARIANS

Index

D

E

F

G

H

R

S

T

U

V

W

Y

Z

Association of Reptilian & Amphibian Veterinarians
5003 SW 41st Blvd
Gainesville FL 32608-4930
USA

ISBN 978-1-5108-5202-0

CURRAN ASSOCIATES INC.
proceedings.com

Annual Conference and Expo of the Association of Avian Veterinarians (AAV 2017)

Soaring to New Heights

Held at ExoticsCon 2017

Washington, DC, USA
29 July – 2 August 2017

Association of
Avian Veterinarians
Advancing & Promoting Avian Medicine and Stewardship